TELEOLOGICAL LANGUAGE
IN THE LIFE SCIENCES

TELEOLOGICAL LANGUAGE IN THE LIFE SCIENCES

Lowell Nissen

ROWMAN & LITTLEFIELD PUBLISHERS, INC.
Lanham • New York • Boulder • Oxford

ROWMAN & LITTLEFIELD PUBLISHERS, INC.

Published in the United States of America
by Rowman & Littlefield Publishers, Inc.
4720 Boston Way, Lanham, Maryland 20706

12 Hid's Copse Road
Cummor Hill, Oxford OX2 9JJ, England

British Library Cataloguing in Publication Information Available

Library of Congress Cataloging-in-Publication Data

Nissen, Lowell A.
    Teleological language in the life sciences / Lowell Nissen.
        p.   cm.
    Includes bibliographical references and index.
    ISBN 0-8476-8693-0 (alk. paper). — ISBN 0-8476-8694-9 (pbk. : alk. paper)
    1. Biology—Philosophy. 2. Teleology. 3. Biology—Language. I. Title.
QH331.N57   1997
570'.1—dc21   97-7424

ISBN 0–8476–8693-0 (cloth : alk. paper)
ISBN 0–8476–8694-9 (pbk. : alk. paper)

Printed in the United States of America

Contents

Preface

Although no definition of teleological language has general acceptance, there is rough agreement that sentences such as "He gathered wood in order to start a fire," "The purpose of digging a well is to find water," and "The function of a hammer is to pound things" are examples. Teleological language is not limited to the explicit use of such expressions as "purpose" and "function." Common expressions such as "flee," "protect," "hide," and "migrate" are also teleological. If an animal flees, for example, it is running in order to escape. There is also general agreement that teleological language has a future orientation, often expressed by saying that, in some way or other, it talks about a later stage to explain an earlier stage.

Aristotle used teleological language to describe the behavior of nonliving things as well as living things, explaining why stones fall and smoke rises in the same terms as why a person gathers wood. Medieval science was heavily influenced by Aristotle, and the use of teleological language continued. With the radical revision of mechanics in the seventeenth century came a general reaction against Aristotelian medieval science, a prominent feature of which was the removal of teleological language. Except for figurative use, it has never been readmitted into the physical sciences. However, teleological language continues in the life sciences to the present day, where its use remains controversial but vigorously defended.

Teleological language is controversial because of some apprehension that it might presuppose either reverse causation or minds. This is not a problem when used in contexts where purpose and intention can be safely assumed, as in much of human psychology and in sociology and anthropology. However, teleological language is also used throughout biology, including physiology and ethology, and shows no sign of abatement, even as physiology merges with biochemistry. It is this continued use which has attracted attention, especially the use of teleological language in describing parts and processes of organisms and the behavior of lower animals and all plants, where minds or adequate minds cannot be safely assumed. The view prevailing to a degree near unanimity among both scientists and philosophers, especially among philosophers of biology, is that, when properly understood, teleological language, even in such contexts, does not presuppose mind or reverse causality. The purpose of this study is to challenge the prevailing view, to present the case conspicuously absent in the debate over the last fifty years—that teleological language is essentially mentalistic.

Although teleological matters have long been a subject of concern, interest picked up with an article by Rosenblueth, Wiener, and Bigelow in 1943 relating teleology to research in automatic control systems. An extended debate followed, cresting in the 1970s but still continuing, receiving impetus of late in the appeal to biological functions in recent theories of meaning and reference and in epistemology. Because the view is so widely held that a nonmentalistic account of teleological language is not only possible but, barring a few details here and there, has already been achieved, considerable effort is devoted to developing the negative case, looking at many analyses and, in the interests of accuracy in reporting the views of others, making extensive use of direct quotation. An alternative account is developed, one based on intentionality, accompanied with a defense explaining how it avoids the defects of its competitors. If the current effort to reduce intentionality to neurophysiology succeeds, then, of course, an intentionalistic account may ultimately still be nonmentalistic; however, if intentionality is essentially involved in teleology, including functions, as here argued, then current efforts to analyze intentionality in terms of biological functions must involve circularity. The general lack of bibliographic tools in the literature prompted the inclusion of a somewhat extended recommended list for the interested reader.

Acknowledgments

The author gratefully acknowledges permission to reprint passages from the following sources.

Clockwork Garden: On the Mechanistic Reduction of Living Things by Roger J. Faber, the University of Massachusetts Press, 1986. Permission granted by publisher.

The Explanation of Behaviour by Charles Taylor, Humanities Press, 1964. Permission granted by author.

"Four Ways of Eliminating Mind from Teleology" by Lowell Nissen, *Studies in History and Philosophy of Science*, 1993:27-48, Pergamon Press Ltd., Headington Hill Hall, Oxford OX3 OBW, UK. Permission granted by publisher.

"A Goal-State Theory of Function Attributions" by Frederick Adams, *Canadian Journal of Philosophy,* 1979, University of Calgary Press. Permission granted by author and publisher.

"Nagel's Self-Regulation Analysis of Teleology" by Lowell Nissen, *The Philosophical Forum*, 1980-81. Permission granted by publisher.

"Natural Functions and Reverse Causation" by Lowell Nissen, in *Current Issues in Teleology*, edited by Nicholas Rescher, Center for the Philosophy of Science, University of Pittsburgh and the University Press of America, 1986. Permission granted by publishers.

The Structure of Science: Problems in the Logic of Scientific Explanation by Ernest Nagel, Harcourt Brace and World, 1961. Permission granted by Sidney and Alexander Nagel.

Teleological Explanations: An Etiological Analysis of Goals and Functions by Larry Wright, University of California Press, 1976. Permission granted by publisher.

"Teleology and the Logical Structure of Function Statements" by William C. Wimsatt, *Studies in History and Philosophy of Science* 3:1-80, 1972. Permission granted by Elsevier Science Ltd., The Boulevard, Langford Lane, Kidlington, Oxford OX5 1GB, UK.

Teleology by Andrew Woodfield, Cambridge University Press, 1976.

Permission granted by author and publisher.

"Wimsatt on Function Statements" by Lowell Nissen, *Studies in History and Philosophy of Science* 8:341-47, Pergamon Press, 1977. Permission granted by publisher.

"Woodfield's Analysis of Teleology" by Lowell Nissen. *Philosophy of Science*, 1983. Permission granted by publisher.

"Wright on Teleological Descriptions of Goal-Directed Behavior" by Lowell Nissen, *Philosophy of Science*, 1983. Permission granted by publisher.

"Wright's Teleological Analysis versus Impossible Goals" by Lowell Nissen, *Proceedings of the Southwestern Philosophical Society. Philosophical Topics* 12, *Supplement*, 1981. Permission granted by publisher.

Chapter 1

Behaviorism

The past fifty years of debate on teleology has produced a great variety of analyses. Some are built on others, and for these, it seems best to group those with similar strategies together. A clearly recognizable strategy is that of analyzing teleological language in terms of overt behavior. During this half-century of debate, the first broad approach to understanding teleological language having general appeal was in terms of the publicly observable behavior of the subject, beginning with Rosenblueth, Wiener, and Bigelow's landmark article, "Behavior, Purpose and Teleology," which appeared in 1943.

Rosenblueth, Wiener, and Bigelow

This article, together with its sequel, "Purposeful and Non-Purposeful Behavior," is a logical place to begin the study of the contemporary teleological debate because after a lengthy period of only sporadic interest the subject took on new life with the article's publication, in considerable part because it also introduced the subject of negative feedback into the debate. Indeed, the authors approach teleology not from a background in the life sciences but from pioneering research in the theory of automatic control mechanisms, of machines that perform tasks hitherto requiring human direction and monitoring. They are confident that the new science of cybernetics would, as a by-product, provide the long-sought key to understanding purposeful behavior in organic life. "Some machines . . . are intrinsically purposeful. A torpedo with a target-seeking mechanism is an example. The term servomechanisms has been coined precisely to designate machines with intrinsic purposeful behavior" (Rosenblueth et al. 1943, 19).

They see their new approach as an extension of the behavioristic movement in psychology. "This essay has two goals. The first is to define the behavioristic study of natural events and to classify behavior. The second is to stress the importance of the concept of purpose" (1943, 18).

> If the term purpose is to have any significance in science, it must be recognizable from the nature of the act, not from the study of or from any speculation on the structure and nature of the acting object. . . . In other words, if the notion of purpose is applicable to living organisms, it is also applicable to non-living entities when they show these same observable traits of behavior. (1950, 323)

Although these essays have been much discussed in the literature on teleology, the authors were unaware, as were most commentators, that the essays contain, not one, but two distinct analyses of purposive behavior. A behavioristic perspective is clearly expressed when the authors say, "The term purposeful is meant to denote that the act or behavior may be interpreted as directed to the attainment of a goal—i.e., to a final condition in which the behaving object reaches a definite correlation in time or in space with respect to another object or event" (1943, 18).[1] This might be called the definite correlation analysis. The other analysis is the one for which the articles became famous: the negative feedback analysis, which speaks of behavior that "is controlled by the margin of error at which the object stands at a given time with reference to a relatively specific goal" (1943, 19).[2] However, since negative feedback has to do with causal pathways and not with behavior patterns, the authors are wrong in construing it as a version of behaviorism, though, the definite correlation analysis does fit behaviorism. Only the behavioristic portion of their work will be considered here. The negative feedback portion will be addressed in chapter 2.

Behavior is passive or active behavior depending on whether the source of energy is within or without the behaving object. "Active behavior is that in which the object is the source of the output energy involved in a given specific reaction. . . . In passive behavior, on the contrary, the object is not the source of energy" (Rosenblueth et al. 1943, 18).[3] A homing torpedo exhibits active behavior; a thrown ball, passive behavior. Although not explicitly noted, the authors apply the definite correlation analysis only to passive systems, as in the following:

> We consider the behavior of a weighted roulette purposeful, much as we deem that the behavior of a magnetic compass deviated from its resting

alignment exhibits purpose. By this we mean that the analysis of the motions of the wheel or of the needle should include the fact that they end in a definite relationship to a specific characteristic in the environment in which they occur. The purposiveness in these instances differs importantly from that recognizable in a servomechanism, however, because the behavior of the latter is active, whereas the behavior of the wheel and that of the magnetized needle are passive. (Rosenblueth et al. 1950, 319)

The negative feedback variety of purposive behavior appears to be restricted to active systems, such as servomechanisms.

The behavioristic definite correlation analysis is excessively inclusive. Nothing is excluded, for everything at all times and in all places has a definite correlation with other objects and events—in fact, with everything else. Richard Taylor, the earliest and most astute critic of these articles, writes, "Indeed, the definition is so broad that it not only fails to distinguish, even in some general way, the feature which it is intended to describe, but makes any behavior whatever, whether active *or* passive, a case of purposiveness" (1950[a], 311). His examples include a clock stopping after running many years, a stone falling from a rooftop and making a dent in the ground, and even dying (1950[a], 311-13).

> At this moment, for instance, I note that the wisps of smoke emitted by my pipe maintain a definite correlation with certain features of the environment, viz., air-currents in the room. Similarly, the passive motion of the rocking chair beside me may, I suppose, be interpreted as directed to a final condition, rest, relative to another object of its environment, the floor. But what, I iterate, is gained by thus calling tobacco pipes, rocking chairs, compass needles, weighted roulettes and so on, *purposeful*? . . . If, as appears to be the case, purposiveness becomes ubiquitous, then the application of the word "purpose" becomes as general as that of "behavior" itself, and, by the criterion suggested, loses its significance simply by having no counterpart. (R. Taylor 1950[b], 330-31)

The authors, in responding to the criticism in Taylor's first of two articles, fail to recognize the awkward consequences of their definite correlation theory: "The time and place at which a clock will stop cannot be considered as a goal or purpose of its motion because this time and place do not influence the running of the clock" (1950, 322). Thus, they reject Taylor's counterexample to the definite correlation theory because it does not meet the criterion of the negative feedback theory, indicating they are not aware of the difference. Further, they accept the examples of a weighted roulette wheel and a magnetic compass as being purposive; yet the time and place they stop do not influence their movements. Unaware that they have offered two distinct analyses of purposive behavior, they appeal to the consequences of the first one, then the other, to respond to Taylor. Clearly, Taylor has not been answered.

A second defect that Taylor notes is that, in order for an object to move toward a definite correlation with something else, that something else must exist.

> Consider, for example, a man groping about in the dark for matches which are not there, but which he erroneously believes to be near at hand, or another who goes to the refrigerator seeking an apple which he mistakenly believes to be there. Now we can say, if we like, that such behavior sequences are incomplete, or that they are clumsy, but they are assuredly purposeful, i.e., directed towards goals, despite the fact that these goals are not achieved and do not, in fact, exist. (R. Taylor 1950[b], 329)

The definite correlation analysis requires that the goal, since it is one part of the items related, exist. However, as the above examples indicate, purposeful behavior has no such requirement.

Consistent with behaviorism, the authors see purpose as something entirely within what is publicly observable: "If the term purpose is to have any significance in science, it must be recognizable from the nature of the act" (1950, 323). They offer as evidence the example of a car following a man down a road with the purpose of running him down and claim that the proper understanding of the behavior is not affected by whether the car is driven by a person or by mechanical sensors and controllers (1950, 319). The example illustrates a pervasive defect of behaviorism—that the same behavior may result from diverse purposes or even none at all. Taylor counters:

> Surely the observable behavior of the car and its driver might be exactly the same, whether the purpose is, as supposed, to overrun a pedestrian, or merely, as a joke, to frighten him, or, indeed, to rid the car of a bee, the driver being in this case wholly unaware that his car is endangering another person. (1950[b], 328)

If various purposes or even no purpose at all may have the same overt behavior, then purpose cannot be defined in terms of behavior nor can behavior supply the complete criteria for purpose. Taylor makes the point with enduring effect by noting that a torpedo pulled along behind a ship by a cable may exhibit the same behavior as an ordinary torpedo; hence, the decision to call the ordinary torpedo purposive and the towed torpedo nonpurposive cannot rest on behavior alone (1950[a], 317). Eve-Marie Engels (1982, 169) argues that the authors make many distinctions that cannot be managed on the basis of observation of behavior alone, noting that not even separating active and passive systems can be accomplished behaviorally. In a well-known passage, William Wimsatt (1972, 20-21) levels a similar criticism against behavioristic analyses by considering a servomechanism sensitive to a certain variety of radiation. Its behavior is that of moving to various sources of this radiation and recharging itself at each for a certain length of time. One source, however, is so strong that it overcharges the power supply and destroys it. Wimsatt notes that the behavior is compatible with the purpose being that of wandering about recharging its power supply, its destruction being an unfortunate accident, but is also consistent with the purpose being that of seeking self-destruction. Because the same behavior can have different purposes, Wimsatt rejects behaviorism and treats purpose as a separate category in his analysis of function statements (1972, 32).

Braithwaite

Richard Braithwaite, in his classic *Scientific Explanation*, divides teleological behavior into two distinct kinds. One kind involves intention and clearly is not behavioristic. "Now there is one type of teleological explanation in which the reference to the future presents no difficulty, namely, explanations of an intentional human action in terms of a goal to the attainment of which the action is a means" (Braithwaite 1964, 324). To avoid problems about the future determining the past, he attributes causal efficacy to intentions. "Teleological explanations of intentional goal-directed activities are always understood as reducible to causal explanations with intentions as causes" (1964, 324-25).

The second analysis and the one for which he is known does not refer to intention and is behavioristic. It is applicable to goal-directed activities of those subjects where attributing intentionality is either questionable or clearly inappropriate, thus underwriting teleological explanations throughout biology.

> I believe that we can go on the orthodox assumption that every
> biological event is physico-chemically determined, and yet find an

important place in biology for such explanations. So what I propose to do is to try to give an account of the nature of teleological explanations which will resolve the philosophical difficulty about the apparent determination of the present by the future without either contravening the usual determination principles of science or reducing all biological laws to those of chemistry and physics. (Braithwaite 1964, 327-28)

Braithwaite recognizes that goal-directed behavior cannot be determined by some peculiar feature of the terminating event. Following E. S. Russell (1945, 144), he takes persistence in the face of obstacles in moving toward a goal as the determining factor.

Coming to a definite end or terminus is not *per se* distinctive of directive activity, for inorganic processes also move towards a natural terminus. . . . What *is* distinctive is the active persistence of directive activity towards its goal, the use of alternative means towards the same end, the achievement of results in the face of difficulties. (Braithwaite 1964, 329)

His analysis is expressed in semiformal notation: "b" stands for system, "c" for causal chain, "e" for initial state of the system, "f" for field conditions, "γ" for the class of causal chains that end in event of type Γ, and "ϕ" for the class of field conditions that determine those causal chains that end in Γ, that is, which are members of γ. A system, b, is plastic if ϕ has two or more members, that is, if there are two or more sets of field conditions, f, which, given initial state, e, of the system, b, determine one or more causal chains, c, which end in Γ (Braithwaite 1964, 329-31).

Although a plastic system usually has more than one causal chain, Braithwaite specifies that plasticity is a function of the field condition, f, not a function of the causal chains. "Nevertheless the essential feature, as I see it, about plasticity of behaviour is that the goal can be attained under a variety of circumstances, not that it can be attained by a variety of means" (1964, 331-32).

An example offered of plastic behavior with multiple causal chains is that of an animal moving to get food. There are several possible causal chains because it may run or walk or creep; it may move along any of several paths; it may move rapidly or slowly; it may move silently or noisily. An example offered of plastic behavior with a single causal chain is that of temperature homeostasis in warm-blooded animals. The causal chain is given as the series of body temperatures. As environmental conditions change, there will be "changes in the activities of the animal (both changes in its total behaviour, e.g., its feeding and migration habits, and changes in its parts, e.g., its sweat glands), yet these changes will be such as to compensate for the changed environmental conditions so that the animal's body temperature does not vary" (Braithwaite 1964, 331).

Behavior is goal-directed, it is said, if in two or more sets of field conditions there are one or more causal chains that are Γ-producing. To be Γ-producing, the causal chain must end in Γ, must reach the goal. No provision is made for failing to realize the goal state. Braithwaite's analysis is designed to account for trial and success, but not for trial and error, sometimes called the problem of goal failure. It ignores the large amount of behavior directed to goals never reached. Israel Scheffler, in a careful analysis of Braithwaite's analysis, remarks, "If Fido, trapped in a cave-in, is in fact never reached, is it therefore false that he pawed at the door in order to be let out? If . . . the psychologist stops replacing the consumed pellets with new ones, is it false to say that the rat continues to depress the lever in order to obtain a pellet?" (1963, 119).

Scheffler considers whether the theory could be enabled to handle goal-failure by deleting the requirement that the causal chains actually achieve the goal. A new difficulty then emerges, for now there would be no way to connect the causal chain to the correct event or condition that constitutes the goal. Scheffler calls this the problem of multiple goals, a designation that has become generally adopted. Suppose, he says, in his familiar example, a cat is crouching before a mouse-hole. If a mouse were to appear, a causal chain would develop beginning with the crouching cat and ending with the caught mouse. However, if a bowl of cream were placed near the crouching cat, a different causal chain would develop, one beginning with the crouching cat and ending with the consumed cream. In each case, the behavior is, let us suppose, plastic.

> It should therefore be a matter of complete indifference, so far as the present proposal is concerned, whether we describe the cat as crouching before the mouse-hole in order to catch a mouse or as crouching before the mouse-hole in order to get some cream. The fact that we reject the latter teleological description while accepting the former is a fact that the present proposal cannot explain. (Scheffler 1963, 120-21)

The problem outlined is the pervasive problem of behaviorism already noted—that the same behavior is compatible with more than one goal, that behavior underdetermines goals—in the preceding section in the example of a car bearing down on a pedestrian being compatible with the purpose being to run down the pedestrian, merely to scare him, or to rid the car of a bee.

Andrew Woodfield believes Braithwaite's goal-failure problem can be resolved by casting the criterion of goal-directed behavior in conditional language. "The difficulty of goal-failure can now be accommodated by invoking counterfactual conditionals of the form 'If obstacles o_x had not been present in field-conditions f_x, which otherwise fall within the variancy, or if an essential member of the set f_x had not been absent (we may call this an obstacle too), the behaviour would have ended in Γ'" (Woodfield 1976, 49). On this view, Scheffler's barking dog would be recognized as goal-directed because if the door had been opened, the dog would have left the cave. This will not do, however, because if contrary-to-fact circumstances are allowed, the dog might do any number of things it was not previously its goal to do. If food were placed before it, it might very well stop barking and eat. That does not mean that it had been barking in order to get food. If an escape hole were dug but the dog were now too weak to move, that would not mean his earlier barking was not in order to escape.

The underdeterminacy of plasticity is apparent in that the plastic behavior may terminate in conditions that are rarely, if ever, goals. Exhaustion is produced under an immense variety of field conditions and in many ways, such as exercise, illness, lack of food, lack of sleep, and loss of blood. Braithwaite's analysis would justify saying, incorrectly, that the parent who stayed up all night tending a sick child did so in order to become exhausted. Hugh Lehman describes a deepwater fish raised to shallow waters by an upward current until it eventually bursts. "According to Braithwaite's analysis of the meaning of function statements, we are allowed to say in this case, the function of the fish's expanding is to produce bursting. But this 'function statement' is unacceptable" (1965[a], 5). Jonathan Bennett observes that "every animal is tremendously plastic in respect of becoming dead: throw up what obstacles you may, and death will still be achieved. Yet animals seldom have their deaths as a goal" (1976, 45). It is because such considerations that several later writers link goal-directed behavior to benefit.

Scheffler also notes that if behavior is plastic in respect to an event that terminates a causal chain, it will also be plastic in respect to any event that uniformly precedes or succeeds Γ (1963, 121). A rat running a maze is plastic in respect to eating the food. That same behavior is also plastic in respect to touching the food, for touching always precedes eating, yet, we say that the behavior is goal-directed toward the eating but not toward the touching.

Woodfield (1976, 42) contests this criticism on the grounds that further experiments utilizing the plasticity criterion could be performed that would eliminate touching the food as a goal. He suggests an experiment involving inserting food through a gastric fistula, presumably involving touching the food without ingesting it. The fact remains, however, that in the absence of such additional experimentation, Braithwaite's criterion for goal-directed behavior would be met and one must, following that criterion, declare, incorrectly, that the behavior is directed to touching the food. Furthermore, each time a new differentiating experiment is performed supporting a revised analysis, a prior event in the causal chain, such as being one centimeter from the food, could always be found, reinstating the problem. Braithwaite's analysis could, of course, be revised to capture Woodfield's observation by expressing it in dispositional form, but, besides not resolving the problem, it would make teleological statements based on one trial more difficult than they in fact are.

Braithwaite says behavior is plastic if there are two or more sets of field conditions that determine the behavioral causal chain. In the example of an animal moving toward food, the successive movements are causally linked not merely to previous movements, forming a chain, but also to constantly changing field conditions, forming collateral linkage. A given step forward is causally linked not only to the previous step forward, but also to the noise coming from the bushes, to a stone rolling down onto the path, to the prey veering to the left, and so on. The movements of an animal, which Braithwaite offers as an example of a causal chain, do not form a series causally separate from the field conditions. The point can be made again with the temperature example. The causal chain, Braithwaite says, is the "chain of body temperatures throughout a period of time"(1964, 331). However, by his own account, environmental changes, such as changes in temperature and availability of food, cause changes in the activities of the animals, such as their running, eating, panting or sweating, which, in turn, contribute to the stable series of body temperatures. The temperatures form a series and the temperatures are caused, but they do not form a causal series, for the causal connections are intricately branched. Requiring there to be a series of events causally linked to each other in the pattern of a chain, more or less separated from lateral causal links, does not fit most goal-directed behavior, such as running to secure food, fleeing a predator, building a nest, or even maintaining stable temperature.

Because minute variation in field conditions is enough to render behavior plastic, hence, goal-directed, the theory can easily be trivialized. Woodfield writes,

> If minute differences in the environment entail differences in field-conditions, then the variancy of the conditions under which, say, a boulder rolls downhill will provide overwhelming evidence of plasticity. But boulders do not *seek* the bottom of the hill. Obviously not all features of the environment are to count as relevant field-conditions. . . . Every causal regularity can tolerate some variation in the conditions under which it operates. (1976, 44-45)

Bennett also notices this problem: "But how do we count possible states of affairs? This bird will eat that worm; and it would still do so if the temperature were .005° warmer, or if there were two more leaves on that tree or five fewer dandelions on the lawn"(1976, 45).

One might suppose the problem to be merely that Braithwaite did not bother to provide criteria separating field conditions that were relevantly different from those that were not. However, it seems an impossible task to find noncircular criteria so exquisitely sensitive as to render relevant, say, variations in season for birds migrating south but not for clouds moving south or variations in ground cover for quail foraging but not for snow falling.

Varying field conditions are not even necessary for goal-directed behavior. We confidently describe a fish patrolling its confining net or a bird flying against the porch screen as trying to get out even though field conditions are stable. Persistence without variation in field conditions is frequently sufficient for accurately determining goal-directed behavior.

Persistence, itself, is not necessary. Suppose a deer, head down and snorting, with no obstacles, runs at top speed toward a hiker. With no obstacles, there is no exhibition of persistence. Yet the hiker judges the behavior to be goal-directed. It is true that the behavior might have shown persistence had obstacles been present, but the hiker, lacking additional information, still judges the behavior as goal-directed and does so with high probability of accuracy. Even if an obstacle, such as a shallow stream that could easily be crossed, halted the deer, revealing an absence of persistence, the hiker would still judge the deer to have charged, its running to have been goal-directed.

Several observations are required in order to determine plasticity, implying that describing the behavior as goal-directed is never justified on the basis of observing a single case. This condition is too restrictive.

The hiker does not wait for two or more trials before judging that the deer was charging. Indeed, he does not even observe one complete trial but judges on the basis of only the beginning of the causal chain. Any successful theory of goal-directed behavior must accommodate the fact that not only humans, but animals as well, routinely make instant but highly accurate judgments concerning which actions are, and which are not, goal-directed based on the barest beginnings of a single trial. Survival depends on using criteria that are far more efficient than Braithwaite's.

Stability of any kind of property in the face of changing field conditions meets the theory's criteria. A large, deeply embedded rock maintains stability of location in the face of a great range of field conditions, including changing temperature and seasons, extremes in weather, even earthquakes. Similar examples can be constructed about stability in color, structure, shape, etc., making the theory far too indiscriminating. Of course, this kind of counterexample could be prevented by making variation in behavior a part of the analysis as well as variation in field.

That would not, however, help when the behavior is that of dynamic systems. If the motion of an animal securing food is plastic because the food is obtained under a variety of field conditions, the movement of water flowing downhill should also be regarded as plastic because the water reaches the bottom under a variety of field conditions. In the case of an animal, some changes in field conditions cause different causal chains, as when a large boulder rolls onto its path and the animal veers around the boulder, while some changes in field conditions do not produce different causal chains, as when the weather changes from overcast to clear and the animal's path in unaltered. Similarly, some changes in field conditions of water flowing downhill cause different causal chains, as when a large boulder rolls into the path and the water veers around the boulder, while some changes in field conditions do not produce different causal chains, again, as when the weather changes from overcast to clear.

There is no doubt that one of the striking features of much goal-directed behavior is that it exhibits plasticity, that the goal is reached under varying field conditions. Nevertheless, much non-goal-directed behavior, especially of the inorganic world, also exhibits this feature, and much goal-directed behavior does not. As Woodfield observes, "Indeed, in the case of human beings, it is a familiar fact that goal-directed behaviour is not always plastic. Sometimes a man won't act unless the conditions are exactly right. The same goes for animals" (1976, 98-99).

Nagel

One of the best-known analyses of teleology and one of those having the largest secondary literature is that of Ernest Nagel. It contains, not one, but two, distinct analyses. The first one, a preliminary necessary condition or requirement analysis, leads to and becomes a part of his major effort, a refined version of a plasticity or covariation behaviorist theory.

Teleological or functional explanations, Nagel says, "can be reformulated, without loss of content, to take the form of nonteleological ones, so that in an important sense teleological and nonteleological explanations are equivalent."

> More generally, a teleological statement of the form "The function of *A* in a system *S* with organization *C* is to enable *S* in environment *E* to engage in process *P*" can be formulated more explicitly by: Every system *S* with organization *C* and in environment *E* engages in process *P*; if *S* with organization *C* and in environment *E* does not have *A*, then *S* does not engage in *P*; hence, *S* with organization *C* must have *A*. (Nagel 1961, 403)

For convenience, this might be dubbed the "necessary condition" analysis, for, ignoring certain refinements, the analysis says that "the function of A in system S is P" means that S produces P and A is necessary for S to do so. For example, the teleological statement "The function of chlorophyll in plants is to enable plants to . . . form starch from carbon dioxide and water in the presence of sunlight" can be reformulated as "when a plant is provided with water, carbon dioxide, and sunlight . . . it manufactures starch . . . only if the plant contains chlorophyll" (Nagel 1961, 403).

Nagel worries whether chlorophyll really is necessary for plants to manufacture starch. Although it is, he says, "logically" or "abstractly" possible (by which he apparently means that the negation is not self-contradictory) and even "physically" possible for organisms to live without chlorophyll, it is not possible, given the organization they have: "there appears to be no evidence whatever that in view of the limited capacities green plants possess as a consequence of their *actual* mode of organization, these organisms can live without chlorophyll" (1961, 404).

It is difficult to assess this response because of uncertainty concerning the meaning of the expression "actual mode of organization." If the phrase is to be taken broadly, as referring, for example, to the having of roots and stems, then chlorophyll is not necessary. If taken narrowly so that it includes reference to making starch in just the way green plants do, saying chlorophyll is necessary for the production of starch is true, but it is also trivial. Lehman observes,

> Suppose that some adventurer discovered a green plant which produces starch but which does not contain chlorophyll. Would we then be forced to withdraw the statement that the function of chlorophyll in green plants is to enable the plant to perform photosynthesis? I do not think so. (1965[a], 7)

Had the example of a function statement been "the function of chlorophyll in plants is to manufacture carbohydrates," the necessary condition analysis would have been less persuasive, for fungi manufacture cellulose, a form of carbohydrate, and fungi have no chlorophyll.

Since organisms meet the demands of life in many different ways, it seems unlikely that the actual methods used are necessary. Some animals escape predation by counterattack, some by camouflage, some by mimicry, some by flight, some by simulating death, some by burrowing. It is only in the sense that allows no significant changes in the organism that one can say that quills on a porcupine are necessary to prevent predation. This approach, which saves the necessary condition claim by disallowing variation, trivializes that position much as in asserting it is necessary to pass through Cleveland in driving from New York to Chicago if all other routes are disallowed.

The failure of a necessary condition analysis of functions is especially apparent in the case of redundant functions. One of the four commonly recognized functions of the spleen is that of producing lymphocytes that protect the body against foreign substances. The spleen is not, however, necessary for the production of lymphocytes, for other parts of the lymphatic system also produce lymphocytes, and a human with the spleen removed usually suffers no ill effects. Vivian Shelanski makes a similar point:

> It is true to say that "Bone marrow functions to produce blood cells", but it is false to assert either that "Bone marrow is necessary for the production of blood cells" or that "Blood cells are produced only if bone marrow is present". For if the bone marrow is destroyed or diseased, the spleen and liver take over the function of producing blood cells in the body. Here we have a case in which the supposedly correlative

statements "The function of X is Y" and "Y only if X" are clearly *not* equivalent. (1973, 399)[4]

Surprisingly, Nagel, as Lehman notes, acknowledges elsewhere that something may have a function even though it is not necessary for the effect to occur.

> Thus, one of the functions of the thyroid glands in the human body is to help preserve the internal temperature of the organism. However, this is also one of the functions of the adrenal glands, so that in this respect there are at least two organs in the body that perform (or are capable of performing) a similar function. Accordingly, although the maintenance of a steady internal temperature may be indispensable for the survival of human organisms, it would be an obvious blunder to conclude that since the thyroid glands contribute to this maintenance they are for this reason indispensable for the continuance of human life. (Nagel 1961, 533)[5]

Thus, although Nagel expresses confidence in the necessary condition analysis, saying, regarding the chlorophyll example,

> If this example is taken as a paradigm, it seems that, when a function is ascribed to a constituent element in an organism, the content of the teleological statement is fully conveyed by another statement that is not explicitly teleological and that simply asserts a necessary (or possibly a necessary and sufficient) condition for the occurrence of a certain trait or activity of the organism, (1961, 405)

by his own account, that analysis will not do.

Nagel goes on to say, as if by way of filling out the necessary condition analysis, that a teleological statement is different from an ordinary causal statement in that it indicates consequences rather than causes. "The difference between a teleological explanation and its equivalent non-teleological formulation is thus comparable to the difference between saying that Y is an effect of X, and saying that X is a cause or condition of Y. In brief, the difference is one of selective attention, rather than of asserted content" (Nagel 1961, 405). According to this, saying "The function of chlorophyll in plants is to manufacture starch" is different from "Chlorophyll in plants is a cause of the manufacture of starch" only in what is emphasized, not in what is said, and is, therefore, equivalent in content (not merely comparable) to "The effect of chlorophyll in plants is to manufacture starch."

Although attractive in its simplicity, this view, which might be called the causal consequence view of teleological language, is neither equivalent to nor implied by the necessary condition view. Something can be a necessary condition without being a causal consequence or an effect, and something can be a causal consequence without being a necessary condition. The breaking of a vase is a causal consequence of its being bumped, but the breaking is not a necessary condition for its being bumped. Nagel is probably influenced by the fact that in conditional statements, the consequent is taken as describing a necessary condition for that described by the antecedent. However, the causal relation is not adequately captured by the conditional. Furthermore, the necessary condition that Nagel needs is a nonsubsequent necessary condition, whereas most consequents of conditional statements describe subsequent necessary conditions.

Engels raises the interesting objection that such a causal consequence analysis of teleology must fail because it does not capture the means-end status of the elements described teleologically; it does not give their role. She sees the kind of causation that, for example, links kidneys with constant chemical composition of the blood as significantly different from the kind that links a storm with the uprooting of a tree, and regards Nagel's causal consequence analysis as trivial because the means-end aspect is ignored (1978, 231).[6]

Nagel is aware that more is needed, for he notes that were the necessary condition analysis correct, there should be equivalence between a teleological statement and its necessary condition formulation and, if there were such equivalence, each should imply the other. However, assuming his necessary condition analysis, although a teleological statement does imply its corresponding nonteleological statement, the reverse often does not. For example, the translation of Boyle's law stating that the volume of a gas at constant temperature varies inversely with pressure "would presumably be 'The function of varying pressure in a gas at constant temperature is to produce an inversely varying volume of the gas,' or perhaps 'Every gas at constant temperature under a variable pressure alters its volume in order to keep the product of the pressure and the volume constant'" (Nagel 1961, 406). Since such a translation would be preposterous, the necessary condition analysis is threatened.

In response to this difficulty, Nagel modifies the necessary condition analysis by introducing the concept of a goal-directed system. "However, the discussion thus far can be accused, with some justice, of naïveté if not of irrelevance, on the ground that it has ignored completely the fundamental point, namely, the 'goal-directed' character of organic systems" (1961, 408).[7] Following Gerd Sommerhoff (1950), he takes as the distinguishing marks of goal-directed behavior the property of plasticity, by which he means that the goal can be reached by more than one path, and the property of persistence, that, when one path is blocked, an alternative is used. It is here that his analysis becomes a variety of behaviorism. Goal-directed systems are such

> that they continue to manifest a certain state or property G (or that they exhibit a persistence of development "in the direction" of attaining G) in the face of a relatively extensive class of changes in their external environments or in some of their internal parts—changes which, if not compensated for by internal modification in the system, would result in the disappearance of G (or in an altered direction of development of the systems). (Nagel 1961, 411)

Nagel's idea is that a teleological statement presupposes a goal-directed system and when that presupposition is satisfied, then a teleological statement is equivalent to its corresponding necessary condition statement.

He illustrates this analysis of a goal-directed system with the example of temperature homeostasis in the human body (1961, 411-17). Let "S" represent the system, in this case, the body, and "A," "B," and "C" represent the diameter of peripheral blood vessels, the secretion of the thyroid gland, and the secretion of the adrenal gland. A, B, and C have certain physiological limits and, in addition, are independent of each other at any given time, a condition Nagel adds to avoid redundant state variables. "G" stands for a certain property of S—in the example of temperature homeostasis in humans, that of maintaining a temperature between 97 and 99 degrees Fahrenheit. Suppose S has property G, along with certain values for A, B, and C, and suppose also that a change in any of those variables would, if no compensating changes occurred, remove G. Under these conditions, if one of the variables assumed a new value and if this were followed by compensating changes in one or both of the remaining variables with the result that S retained or regained G, then S would be considered a directively organized system. In the example at hand, if the adrenalin increased significantly and if, say, the peripheral blood vessels also increased in size, with the result that the body stayed within the normal temperature range, then the body would be a directively organized system. Nagel later expands the analysis to include factors outside the organism by adding external variables.

Nagel's analysis of goal-directed systems appears made to order for the many homeostatic processes of organisms, such as glucose regulation, potassium and sodium regulation, maintaining the sleep-waking cycle, and so on. It can be extended to cover activities of the whole organism, such as pursuit, flight, and migration, but only awkwardly. In the case of pursuit, in order to find something that is maintained by compensating adjustments when disturbed, G seems to have to be the moving toward the pursued rather than the reaching of it.

To show that his analysis is properly discriminating, he considers a pendulum disturbed by a gust of wind (1961, 419-420). Suppose the state of rest of the bob at its lowest point to be the goal state and that a gust of wind set the bob in motion. The oscillations gradually decrease in amplitude, and eventually the bob returns to its initial position of rest. It would appear that pendulums "continue to manifest a certain state or property G (or . . . exhibit a persistence of development 'in the direction' of attaining G) in the face of . . . changes in their external environments." If the analysis were to fit pendulums, it clearly would be too broad to be of any value in explicating teleological statements, for we do not consider the oscillations of pendulums goal-directed behavior.

Nagel's response is to call attention to his requirement that the state variables be mutually independent at any given moment: "the possible values of one state variable *at a given time* will be assumed to be independent of the possible values of the other state variables *at that time*" (1961, 412).[8] The disturbing force and restoring force do not meet this requirement, for by the use of appropriate equations, each force at a certain instant can be calculated from the other at that same instant.

Woodfield argues that Nagel's independence requirement would rule out genuine goal-directed systems, as well. "In Nagel's homeostat, whenever A varies as a result of a change in S, B and C vary by a corresponding amount in such a way as to keep the temperature constant; *therefore* A, B, and C are not functionally independent" (1976, 69). Nagel, in his 1977 lecture, "Teleology Revisited," acknowledges the problem. "On pain of contradiction, this surely *cannot* mean that the variables must be orthogonal in those very circumstances in which the system *actually is* in the goal-directed state relative to a given goal G," and responds, "What the requirement *does* mean is that, apart from those situations in which determinate relations hold between the variables because of their role in goal-directed processes, the known (or assumed) 'laws of nature' impose no restrictions on the simultaneous values of the variables" (1977, 275). He concludes with surprising equanimity that goal

direction is, therefore, relative to the laws currently accepted and that
before Newton, the pendulum "might very well have counted as
goal-directed" (1977, 275).

If Nagel means that before Newton the pendulum might have been
regarded as goal-directed, that would be historically false, but at least
understandable. If, however, he means that before Newton the pendulum
really was goal-directed, that would surely strike most as either incoherent
or embracing a degree of antirealism in science that would certainly be
rejected by physicists, then and now. Making the goal-directedness of
pendulums dependent on culture has the awkward consequence that a
pendulum may really be goal-directed, not merely thought so, in one
culture and, if carried across the border, not be goal-directed in another,
even if the residents spoke the same language. Finally, making
goal-directedness depend on knowledge of laws requires that we often
would not know whether a given behavior is goal-directed in cases where
we clearly do know. We confidently judge that certain behavior is
goal-directed, for example, that a dog is pursuing a squirrel, even though
we do not know the laws involved.

The situation is, however, more serious than either Woodfield or Nagel
suppose. According to the last formulation, Nagel's independence
requirement excepts "those situations in which determinate relations hold
between the variables because of their role in goal-directed processes."
Thus, for example, in a thermostatically controlled heating system, the
lawful relation between ambient temperature and heat production is
excepted from the independence requirement. The independence
requirement, however, is a part of the set of criteria Nagel uses to
determine when a process is goal-directed; therefore, we do not know
prior to applying these criteria whether a given process is goal-directed.
Since we lack prior and independent knowledge of which processes are
goal-directed, it cannot be objectionable to select the disturbing and
restoring forces of an oscillating pendulum as the relevant variables.
Nagel's revised formulation would then exclude these variables from the
independence requirement. Because the disturbing force is considered the
primary variation and the restoring force the adaptive variation, these
forces are beyond the jurisdiction of the independence requirement. Thus,
the pendulum example would now be an instance of a system that Nagel's
criteria declare to be goal-directed but which is not, in fact, so regarded,
even by Nagel. Along with this example, the many equilibrium-restoring
systems, such as vibrating solids, distorted elastic bodies, disturbed
liquids, etc., become counterexamples. On the other hand, if it were

objected that we are not allowed to use disturbing and restoring forces as the primary and adaptive variations because they are not related to genuine goal-directed process, then the criteria could not be applied at all, for one would have to know whether a process were goal-directed before applying the very criteria that determine whether it is goal-directed. Nagel's revised independence requirement seems to work because we all agree before applying the criteria which are the goal-directed systems, making it seem clear, though it really is not, that ambient temperature and heat production are to be excluded in the one, but that disturbing force and restoring force are not to be excluded in the other.

A quite different flaw in Nagel's use of the independence requirement becomes apparent when we recall that, other than cotemporary independence, no restrictions are placed on which properties may be used as values for the state variables. Regarding the independence requirement, Nagel says,

> This assumption must not be misunderstood. It does not assert that the value of a variable at one time is independent of the values of the other variables at some *other* time; it merely stipulates that the value of a variable at some specified instant is not a function of the values of the other variables *at that very same instant.* (1961, 412)

This provision alone would have met Woodfield's objection, because the disturbing variations and compensating variations rarely or never occur at the same instant. Adrenaline increases after the temperature drops, perspiration after it rises.

However, this provision, reasonable as it is, allows too much. Instead of using disturbing and restoring forces, disturbing and restoring motions of the pendulum could be used. Indeed, Nagel elsewhere uses an example of a woodpecker's searching for grubs and has the relative positions of the bird's bill and the larvae as the variables (1977, 276).[9] The pendulum's disturbing and restoring motions are not simultaneous. Because of inertia, the restoring motion occurs after the disturbing motion even though the restoring force is simultaneous with the disturbing force. Since once again the independence requirement is not violated, the pendulum counterexample is reinstated.

There are other difficulties. Nagel's analysis claims that a functional statement is equivalent in meaning to a necessary condition statement in the context of a goal-directed system. Thus, the statement "The function of the heart is the circulation of the blood" is alleged to be equivalent to "The heart is necessary for the circulation of the blood" in the context of a goal-directed system such as the human body. This formulation allows Carl Hempel's well-known heart sounds counterexample (Hempel 1965, 305). Accepting Nagel's use of "necessary," "The heart is necessary for the production of heart sounds" is true, but "The function of the heart is the production of heart sounds" is false. Woodfield notes, "But since Hempel's 'heartbeat' counter-example is still effective against this, a tighter connection must be forged between F and the goal-directedness of S" (1976, 124). This kind of counterexample is possible because Nagel's formulation does not require that the effect contribute to reaching the goal. Peter Achinstein makes the same point when he says, regarding Nagel's analysis, "But this is incomplete as it stands since doing y, on this analysis, need make no contribution to the attainment of G" (1977, 343).[10] As noted earlier, this is one reason why some analysts have included reference to benefit in their analyses.[11]

Francisco Ayala, who generally approves Nagel's analysis, prefers such a modification, suggesting that the analysis be altered so that

> A feature of a system will be teleological in the sense of internal teleology if the feature has utility for the system in which it exists and if such utility explains the presence of the feature in the system. . . . A structure or process of an organism is teleological if it contributes to the reproductive efficiency of the organism itself, and if such contribution accounts for the existence of the structure or process. . . . Chemical buffers, elastic solids and a pendulum at rest are not teleological systems. (1970, 13)

However, for something to be useful there must be something for which it is useful, that is, its use, otherwise called its end, purpose, or goal. The utility Ayala has in mind is reproductive efficiency of the organism. Appealing, whether explicitly or by implication, to the concept of use, goal, or end to explain goals and ends is circular.

Further, what is dangerous to the individual might be useful for the species, as, for example, incubating eggs and feeding and protecting the young. In addition, some behavior is goal-directed but harmful to both the agent and the species, as in self-destructive human behavior. It may be useful in the sense of fulfilling a purpose but not in the sense of being beneficial. Finally, Ayala says nothing about how goals are identified. Although he is confident we can identify the goals of the structures of organisms and of man-made tools and can perceive their absence in elastic solids and pendulums, he does not indicate how this is done. The difficulty is how to do this without being arbitrary or circular. Neither kind of error is easily detected if there is widespread agreement on what the goal is, as is the case with the goal being reproductive efficiency or species survival.

Although self-regulating artifacts, such as thermostatically controlled heating systems, fit Nagel's analysis, non-self-regulating artifacts do not. A hammer has a function even though it is not a part of a self-regulating system or, in fact, a system of any kind. The same can be said of knives, chairs, and pencils; indeed, most artifacts are of this kind.

Many systems are self-regulating without being goal-directed. A watershed drains, say, into five rivers. If an obstruction, such as a beaver dam, diminishes the flow in one, the flow increases in the others. The blocking of the one and the increased flow in the others are not simultaneous, so the independence rule is not violated. If the arrival at sea is considered the goal state, the watershed can, using the language of Nagel's analysis, be said to compensate for the disturbance because the blocked water reaches the sea another way. Accordingly, that analysis requires that a watershed be considered a goal-directed system, though it is not, in fact, so regarded.

We might also consider the atmosphere over a certain area and identify as the goal state the atmosphere's maintaining a certain temperature range. A disturbing influence appears every day with the rising of the sun. The air mass warms and rises; cooler air moves in below, and, as a result, the temperature of that area remains within the goal range. If a building with a thermostatically controlled furnace fits Nagel's criteria for goal-directed systems (Nagel 1961, 418), then a local weather system does as well. Instead of a thermostat's turning off the furnace when the temperature reaches the upper limit, cooler air moves in below. Surely goal direction does not depend on that difference.

If the water level of a pool were controlled by a float valve in the intake channel, the system would doubtlessly be considered goal-directed. Presumably, it would also be considered goal-directed if the level were controlled by installing a drain channel at a certain level. If goal direction were understood as Nagel declares, it should make no difference whether the water level is controlled by a channel formed naturally by erosion or by the bank or shore, making all lakes, ponds, and streams goal-directed systems.

Morton Beckner offers as a counterexample a lake in which is maintained a certain predator-to-prey ratio. "This system is directively organized with respect to the biomass ratio of predator and prey fishes.

But we would not say that a function of the trout is to eat the bluegills, although this does play a role in the regulation of the ratio" (1959, 156).

These examples are only a few in a large class of examples of systems that fit Nagel's account of self-regulative systems but that no one considers goal-directed systems. Since these examples all exhibit primary variation and adaptive variation, it is clear that self-regulation fails to properly identify goal-directed systems.

Not only does Nagel's analysis judge some behavior to be goal-directed that clearly is not, but it rejects as goal-directed some behavior that, though exhibiting no primary or secondary variation, clearly is. Nagel intends his analysis to apply generally: "The definition of directively organized systems has been so stated that it can be used to characterize both biological and nonvital systems" (1961, 418). Consider a person's inserting a key into a lock to open a car door, an action performed countless times before. The key is inserted without fumbling, it fits, the lock opens, etc. The action is performed in a smoothly coordinated manner. There is no disturbing variation, hence no compensatory variation. Nevertheless, the action is goal-directed. Such non-compensating behavior occurs over a wide range of behavioral levels, as illustrated by a dog pawing at the door in order to be let out or a rat pressing a bar to get food. Woodfield's observation, noted earlier, is again appropriate: "Indeed, in the case of human beings, it is a familiar fact that goal-directed behaviour is not always plastic. Sometimes a man won't act unless the conditions are exactly right. The same goes for animals" (Woodfield 1976, 98-99).

It might be objected that Nagel does not require that there actually be a disturbing event, but only that if there were, there would be a compensating one. "To be goal-directed, the process must satisfy the much stronger requirement that *were* the blood inundated with water to a greater or lesser extent than was actually the case, the activity of the kidneys or of the muscles and skin *would* have been appropriately modified" (Nagel 1977, 273). Each of the above examples could be modified to take care of this objection by including the disturbance but removing the compensatory action. The person inserts the key into the lock to open the door, but the key does not operate the lock and, after one unsuccessful effort, the person simply gives up. The dog paws at the door for a moment, the door remains closed, and the dog goes his way. The rat presses the lever, no pellet drops, and he abruptly ceases. In each of these cases, there is a disturbing event but no compensatory behavior. In spite of the lack of this crucial element, we would still consider the behavior goal-directed.

Adaptive behavior can be eliminated not only by removing the disturbance or by having the subject give up, but also by making it impossible for compensatory behavior to occur. If a person were trying to lift a heavy weight and were suddenly subjected to the disturbance of a wrenched muscle, causing him to fail in the attempt and not try again, he surely would not thereby be regarded as a system without goal direction. Indeed, the word "try" seems reserved for just those cases in which it is a goal-directed system that fails.

Teleological systems that fail but still retain their teleological character are not limited to persons or even organisms. Nagel includes mechanical examples among the things that can be self-regulative systems. "Nevertheless, it has been possible to construct physical systems that are self-maintaining and self-regulating in respect to certain of their features, and which therefore resemble living organisms in at least this one important characteristic" (1961, 410). However, a machine designed to be self-regulatory might never work and still be correctly described teleologically. This may be an elementary observation, but it is damaging to the self-regulation theory of goal-directed systems. A defective thermostatically controlled heating system may be described as a mechanism that was "supposed to" keep the temperature within a certain limit. A defective homing torpedo may be described as a device "made to" pursue and destroy a ship. A defective governor-controlled steam engine may be described as one "designed to" maintain its revolutions per minute within a certain range. Phrases such as "supposed to," "made to," and "designed to" are teleological. We do not say, in respect to a watershed, should one of its rivers become blocked, that it was "supposed to," was "made to," or was "designed to" convey water to the sea.

The analysis says that goal-directed systems "continue to manifest a certain state or property G (or that they exhibit a persistence of development 'in the direction' of attaining G)." The expression "in the direction of" accommodates occasional or contingent goal failure. A rabbit fleeing a dog may perform behavior that is "in the direction" of escape even though it is in the end unsuccessful. Nevertheless, much behavior is directed to goals that are impossible and not merely improbable, such as building a perpetual motion machine or, as Douglas Ehring suggests, traveling at velocities greater than that of light.[12] The goal need not be so esoteric. Much escape behavior is futile, the laws of nature being such that, given the slowness and weakness of prey, the quickness and strength of predator, lack of cover and camouflage, escape is physically impossible; yet we confidently judge the organism to be doing what it is doing in order to escape. The impossibility, as Ehring notes, may also be a logical impossibility such as seeking to derive a proposition from premises that logic disallows.

Nagel's account of self-regulation was initially restricted to cases where the disturbing variation is internal to S, for example, where A assumes a value that takes S out of the G-state. Later, he expands the analysis by allowing the disturbance to come from outside S as well.

> We may now relax the assumption that the external environment E has no influence upon S. But in dropping this assumption, we merely complicate the analysis, without introducing anything novel into it. For suppose that there is some factor in E which is causally relevant to the occurrence of G in S, and whose state at any time can be specified by some determinate form of the state variable "Fw." Then the state of the enlarged system S' (consisting of S together with E) which is causally relevant to the occurrence of G in S is specified by some determinate form of the matrix "$(AxByCzFw)$," and the discussion proceeds as before. (Nagel 1961, 417)

This expansion is needed because the analysis would be incomplete if it could not account for, say, an organism's maintaining constant temperature when environmental temperature changes. However, the expansion brings with it new problems.

Where S is an organism, S has a clear boundary. We have no trouble distinguishing trees from their non-tree surroundings. The enlarged system, S', however, has no clear boundaries. In the case of temperature regulation, would the system include the organism and just the surrounding layer of air? If not, how much more would be included? Weather disturbances involve fairly large geographical areas and climatic changes involve immense regions. If the disturbance were a cold wave moving into the Ohio valley, would the entire valley be a part of S'? If the disturbance were the coming of winter in the Northern Hemisphere, would S' now include the entire Northern Hemisphere? If the feeding behavior of shore birds varies with the tides, would S' include the moon? Things actually described teleologically, such as dogs and hearts and gastric juices, simply do not share the boundary vagueness or occasional immense sizes of Nagel's expanded system.

Nagel's enlarged systems are unstable. When the system is an organism, its boundary is fairly stable. Where a purple finch ends and the non-finch surround begins does not vary much or quickly. In contrast, Nagel's expanded system constantly enlarges and contracts as disturbances outside the organism expand and recede, and in all sorts of unpredictable ways. For a prairie dog, at one moment, the system includes a circling hawk, while in the next, it contracts to the walls of the burrow. Such expanded systems change too quickly for teleological questions to arise or teleological judgments to be made.

Apart from the instability of subject matter, there is the problem that expanded systems do not provide the kind of subject matter teleological talk is about. A teleological judgment is, for example, that a coyote is chasing a rabbit. It is the coyote's behavior that is being described teleologically. However, on Nagel's analysis, the system is the coyote and the rabbit, plus the ground cover, the lay of the land, and perhaps the wind and weather. We do not even have a name for such a system. It is not, as we sometimes say, a natural kind. What is claimed to exhibit teleology is the coyote, whereas the expanded analysis makes what exhibits teleology an amorphous, unnatural, and useless unit about which we have no interest.

The enlarged system is not even self-regulating. The organism may exhibit temperature stability as night replaces day or winter replaces summer, but S′, which includes both the organism and the surrounding area, does not. Far from being the modest modification Nagel anticipates when he says, "we merely complicate the analysis, without introducing anything novel into it," his final version renders the analysis inapplicable to teleological behavior involving external stimuli. Nagel summarizes his analysis by saying,

> In a sense, therefore, a teleological explanation does connote more than does its *prima facie* equivalent nonteleological translation. For the former presupposes, while the latter normally does not, that the system under consideration in the explanation is directively organized. Nevertheless, if the above analysis is generally sound, this "surplus meaning" of teleological statements can always be expressed in nonteleological language. (1961, 421)

Engels suggests a reason how his analysis can be maintained in the face of overwhelming defects when she says Nagel's analysis is acceptable only when the example used is one that is independently teleological, allowing the claimed causal relation in just those cases in which it is also a means-end relation.[13]

The reasons why behavioristic analyses fail to capture what is essential about teleological behavior are many and diverse, exhibiting both underdetermination and overdetermination. The same behavior can occur with different goals and the same goal may be present in different behaviors. Behaviorism, in the interests of securing publicly observable subject matter, substitutes evidence for something with that thing.

Notes

1. Eve-Marie Engels (1982, 167) sees this duality simply as ambiguity in the concept of goal: "'goal' bedeutet ja einerseits *'final condition'* in dem bereits angegebenen Sinn, zum anderen ist der Begriff jedoch *instrumental* gedacht als das Objekt, das Signale aussendet und dadurch kontrollierenden Effekt hat, indem es das betreffende Verhalten dirigiert."

2. It may be that the authors espouse even a third view of purposive behavior. A properly balanced roulette wheel, they say, is purposeless because it stops at various numbers randomly, and a weighted roulette wheel exhibits purpose because it does not stop randomly, suggesting that purposeful behavior is simply nonrandom behavior (Rosenblueth et al. 1943, 19; 1950, 319).

3. Rosenblueth et al. are not, however, consistent on whether passive systems can be purposeful. On one hand, all purposive systems are claimed to be active, for the definition of purposive behavior cited earlier occurs in a paragraph describing active behavior, and their summary table, p. 21, lists purposeful behavior only under active behavior. However, in the 1950 article, p. 325, they say, "Passive behavior may be purposeful or non-purposeful, much like active behavior."

4. Morton Beckner (1969, 155) observes,

> Organisms commonly have alternative means of performing the same function. My right kidney excretes urea, but if it is damaged the left one does the job; so my right kidney is not necessary for urea excretion, although that is one of its functions. Sweating aids in temperature regulation; but if I lose the ability to sweat I can make do by panting and suitable choice of behavior.

Lehman (1965[a], 8) lists the kidneys and the parts of the liver and talks about broader redundancy capability, summarizing: "Where there are multiple methods for creating some effect which is necessary for life of the body, then the destruction of one of these without loss of the required effects does not lead to the conclusion that the item served no function."

Robert Cummins (1975, 744-745), in discussing Nagel, adds brain hemispherical redundancy to that of kidneys. Frederick Adams (1979, 504) notes, in examining Nagel's position, "It is not a necessary condition for attributing a function to x that x be uniquely necessary for y. There are many examples of fall-back mechanisms in biology which sharpen this point." He mentions both the thyroid and adrenal glands in temperature regulation and both the kidneys and sweating in removing water.

5. This was noted also by Lehman (1965[a], 9).

6. "Was Nagel hier ausspart, ist die für teleologische Sprache gerade spezifische Thematisierung einer Mittel-Zweck-Beziehung. . . . Denn während niemand behaupten würde, dass es die Funktion des Sturmes ist, Bäume zu entwurzeln, als Mittel für den Zweck der Entwurzelung der Bäume zu fungieren . . . soll durch eine teleologisch erklärende Formulierung die Präsenz von A (Nieren z. B.) in einem System mit seiner Eigenschaft als Mittel zur Erfüllung eines normativen Sollzustandes des Systems oder seiner Teile gerechtfertigt werden. . . . Zweifellos stimmt es, dass in der Ausgangsformulierung (I) der Effekt von A für B ausgedrückt wird. Aber diese Enthüllung scheint mir trivial." She adds, p. 233, "Ein Bedingungsverhältnis ist noch kein Relevanzverhältnis. M.a.W.: dass A die Bedingung für P ist, bedeutet noch nicht, dass es zweckmässig ist, dass A die Bedingung für P ist."

7. Elizabeth Valentine (1988) unaccountably ignores Nagel's transition to an analysis involving self-regulation and treats his necessary condition as the final one.

8. This is repeated and defended in his "Teleology Revisited," pp. 273ff.

9. Indeed, Thomas Simon (1976, 59), regarding Nagel's independence or orthogonality requirement, says, "the orthogonality condition would not exclude simple causal inverses if suitable time delays were introduced. Input-output could always be made orthogonal relative to output-input by simply delaying the latter relative to the former."

10. A. C. Purton (1979, 15) agrees: "Thus we might usefully modify Nagel's account by saying that i has function F in S just if i is necessary for F in S, and F is or contributes to a 'goal' of S."

11. Thus, Richard Sorabji (1964, 292) says, "But the connection with the notion of a good is neglected in a number of the analyses that have recently been offered, e.g., in those of R. B. Braithwaite, and E. Nagel. If this connection is neglected it becomes difficult to explain why we

should not say, to borrow an example from Hempel, 'the function of the heart is to produce a throbbing sound'".

12. Ehring (1984[b]) expands the criticism by noting that Nagel's primary variation is a change that, were a compensatory variation not to occur, would take the system out of G-state. This rules out the possibility of the system being directed to a goal that will be achieved no matter what variations occur; that is, it rules out anything necessary as a possible goal. He offers as a counterexample a person trying not to exceed the velocity of light. Traveling at sub-light velocity surely is a possible goal, but it is disallowed by Nagel. He combines these criticisms regarding impossible and necessary states by saying that Nagel incorrectly requires that all goals must be contingent.

13. "Erst wenn also das betreffende System ohnehin teleologisch strukturiert ist, trifft die Annahme der Äquivalenz teleologischer und nichtteleologischer Erklärungen zu. Denn dann decken sich Ursache-Wirkungszusammenhänge mit Mittel-Zweckzusammenhängen; die Voraussetzung für eine mögliche Übersetzung ist vom Gegenstand her bereits garantiert" (Engels 1978, 234).

Chapter 2

Negative Feedback

Negative feedback theory arose out of engineering theory of control mechanisms. Its application to teleology dates from the 1940s, followed by years of general neglect when interest was directed elsewhere, but it experienced a modest rebirth in the 1980s. Its considerable and enduring appeal is no doubt due in large part to the fact that most organic behavior and most machine behavior that simulates organic behavior utilize feedback. Its wide application in dynamic devices that are able to maintain a steady state in the face of internal or external changes is especially impressive.

Rosenblueth, Wiener, and Bigelow

Rosenblueth, Wiener, and Bigelow, as earlier noted, initiated both the behaviorist definite correlation view, already discussed, and the negative feedback view. Both were subsequently taken up and developed by others.

The authors present negative feedback theory as a behaviorist theory. Wimsatt points out that this is not correct: "Rosenblueth, Wiener, and Bigelow have made it eminently clear that they are not interested in the internal structure of a system, but only in its externally observable behavior. But even the standard diagrams of feedback systems . . . characteristically represent feedback loops as wholly or partially internal features of the system" (1971, 243).

Woodfield also regards negative feedback not as a behavioral concept but a structural one.

> If an observer looks only at the output or end result of a negative feedback system, he will not be able to tell with certainty whether the system has achieved that result by feedback. The steadiness or invariance he sees might have arisen in other ways. . . . The fact that

> there must be two causal processes, occurring in separate media, which
> join to form a closed loop, shows that feedback is a structural notion.
> . . . Rosenblueth *et al.* were confused about this point in their paper,
> partly owing to their operationalist bias. They stated that feedback is an
> observable property of behavior. In fact, feeding back is a process,
> dependent on structural properties of the system. (1976, 186)

Since one cannot determine whether a system operates by negative
feedback merely by observing its behavior, feedback cannot be purely a
behavioral notion.

Although the authors provide no formal definition, they understand
negative feedback behavior as behavior controlled by the margin of error
of the behaving object in relation to the goal by means of a portion of the
output being fed back to the input with the sign reversed.[1] This requires
comparison with the goal, which requires something real and existing as
the goal. "If a goal is to be attained, some signals from the goal are
necessary at some time to direct the behavior" (Rosenblueth et al. 1943,
19). "Purposeful behavior requires that the acting object be coupled with
the goal, that is, that the object register messages from its surroundings"
(Rosenblueth et al. 1950, 324).

Many different control mechanisms clearly do operate by negative
feedback. Some are electrical, as electronic amplifier circuits, some
mechanical, as various kinds of motor governors, some a combination of
both, as the self-guided missiles and thermostatic heating systems; but
many are organic, such as the systems that control sodium and potassium
levels in the blood, insulin and thyroxin secretion, and so on. It surely has
occurred to some that negative feedback analysis and Nagel's covariation
analysis may be complementary rather than competing and that feedback
could serve as the mechanism responsible for homeostasis in an
environment of change, explaining how a later variation manages to
compensate for an earlier disturbance. It would be interesting to see
whether combining these two analyses would better address the
considerable difficulties each faces on its own.

The problem of the unrealized goals, discussed earlier, which caused
difficulty for the definite correlation analysis, also troubles the negative
feedback account. Because a feedback analysis requires the goal to exist,
it is unable to handle the common experience of searching for something
that is not there, the most frequent variety of which must be looking for
food when there is none. The fruitless search is certainly goal-directed,
but there is no appropriate object to emit the reference signals needed to

generate the error calculation. Engels points out that examples of keeping phenomena, such as a thermostatic system's maintaining a certain temperature, make the problem of the unrealized goal starkly evident. The former temperature certainly does not send a goal signal, neither does the temperature not yet reached. Indeed, the signal does not come from the goal-object at all, but from the setting of the thermostat.[2] Later advocates of negative feedback have responded to such criticisms by claiming that certain states of the system, such as the thermostat's setting, are *signs* of the goal-state, with the signals coming from the signs, not the goal. This interesting idea will be discussed later.

Scheffler identifies an important ambiguity in the authors' description of goals. They describe a goal both as something that sends signals to the behaving object and as a correlation between the behaving object and another object or event. Very likely the two ways of understanding purposeful behavior, as feedback and as definite correlation, have their origin in this ambiguity. In their example of a torpedo guided toward a ship, the ship is the goal in the first sense and contact between the torpedo and the ship is the goal in the second. Scheffler suggests calling the first kind of goal the goal-object (1963, 113). What Taylor identified as the problem of the missing goal is, therefore, more accurately described as the problem of the missing goal-object.

Negative feedback is very common in goal-directed behavior, but hardly sufficient for it. Radios and televisions, Woodfield observes, contain negative feedback circuits but are not goal-directed (1976, 189). Neither, he says, is negative feedback necessary for goal-directedness, for feedback is a process extending over time and cannot operate on instantaneous behavior; but some goal-directed behavior occurs instantaneously. "For example, I take it that when a cuttlefish ejects its tentacles in order to catch a shrimp, it performs a goal-directed response. But this response is not controlled by feedback" (1976, 191). Nor is it necessary for continuous goal-directed behavior. "For example, an animal caught in a trap may snap and bite haphazardly in order to escape; a man may shout or yell in order to attract attention, and so on."[3] The authors, themselves, furnish an example.

> Similarly, a snake may strike at a frog, or a frog at a fly, with no visual or other report from the prey after the movement has started. Indeed, the movement is in these cases so fast that it is not likely that nerve impulses would have time to arise at the retina, travel to the central nervous system and set up further impulses which would reach the

muscles in time to modify the movement effectively. (Rosenblueth et al. 1943, 20)

They regard such behavior as not purposiveful. However, Bennett, who takes a position similar to that of Woodfield, comments on this passage: "Much behavior which does not meet this condition, and so does not involve 'feed-back' in Wiener's sense, would still be counted as teleological by Nagel and Taylor and, I suggest, by common sense" (1976, 61).

As Woodfield sees it, feedback explanations have limited value. They are useful in explaining the means by which a goal is reached (1976, 192) but not in explaining the nature of goals or goal-directed behavior, for they take goals for granted. "It would appear, then, that only careless and superficial thinking would lead a philosopher to claim that the concept of feedback control wholly or partially *explicates* the concept of goal-directedness" (1976, 193).

Rosenblueth et al. consider a servomechanism as a prime example of a machine capable of goal-directed behavior, but even this is subject to dispute. Taylor sees this as a simple misuse of analogy: "The expression 'target-seeking missile' is, in fact, metaphorical" (1950[a], 316). Woodfield makes a similar assessment: "When we describe servomechanisms as goal-directed we are employing a mentalistic analogy" (1976, 194). Their goal-directed behavior "consists in the fact that they behave as if they had desires and beliefs in virtue of the fact that they are feedback systems" (1976, 193). It does nothing, however, for the quality of the teleological debate that advocates and critics alike ignore the fact that servomechanisms are also artifacts, and artifacts have purposes quite independently of how they operate. When a system, in addition to being an artifact, exhibits behavior tending toward a certain state, caution is in order so that purpose is not attributed to it for the wrong reason.

Throughout their two essays, the authors display a surprising insensitivity to the various forms teleological language can take, making error in the analysis of purpose all but inevitable. Thus, they say that a clock does not exhibit purposive behavior because "there is no specific final condition toward which the movement of the clock strives" (1943, 19). However, "striving," itself, is a purposive concept, so the explanation of why the clock does or does not exhibit purposive behavior accomplishes nothing. In the debate between the authors and Taylor concerning whether a radar-controlled gun exhibits purpose, they say that "if the gun seeks the car of the commanding officer of the post, as this officer drives by, and destroys it, surely the purpose of the gun differs from that of the designer" (1950, 318). Well, if the gun really did seek the commanding officer's car, its behavior would no doubt exhibit purpose. This would be so, however, not because of something about the nature of purposive behavior, but merely because to seek something is to engage in purposive behavior toward it. Using unrecognized teleological terms in the analysis of teleological language, enabling one to produce impressive results with little argument, has plagued teleological discussions since ancient times and continues unabated.

Adams

After vigorous criticism, understanding teleological language in terms of negative feedback gave way to several other kinds of analyses, but when these also ran into difficulties, it seemed reasonable, after three decades, to take another look at feedback. Frederick Adams, well aware of the earlier debate, published in 1979 an improved version, the most important feature of which was replacing the goal as the source of signals to be fed back by a representation of the goal.

He emphasizes the connection of the concept of a function with that of a means-end relation: "Let me begin by suggesting that the essential feature of a functional relationship is that of a *means-end* relation. For a structure x to have a function y is, essentially, for x to do y in a system S and for y to lead to the fact that the system is able to output a value O" (1979, 494).

However, he does not distinguish means-end relations from causal relations. "Purely causal relations are sometimes thought of as means-ends relations. When a variable volume gas is heated it expands. Its expanding is the means by which the gas maintains a constant temperature" (1979, 495). The expansion of a heated gas both causes its temperature stability and is a means to that end. The moon causes tidal movement, and Adams suggests it also is a means to achieving tidal movement.

Because means-end relations are, for Adams, simply causal relations, the analysis needs further restrictions, for not all effects are functions. Circulating the blood and producing heart sounds are both effects of the heart, but only the former is a function. Hempel made the division on the basis of what contributed to the well-being of the agent. However, this fails to exclude fortuitous beneficial effects, which, according to Adams' explanation of the example of a soldier's Bible stopping a bullet, are effects that are beneficial but not intended (Adams 1979, 500). Not only may effects be beneficial to the agent without being functions, but effects may be functions without being beneficial. The function of salmon swimming upstream is to spawn even though it does not benefit the swimmers. Some people build devices having functions that are destructive to the builders (Adams 1979, 502). Therefore, fortuitous effects cannot be excluded by attaching a good consequence clause to the analysis.

Adams' solution, reminiscent of that of Nagel discussed earlier, is to require that a means-end relation is a functional relation only if it occurs within a goal-directed system.

> So, e.g., we can avoid attributing to the moon the function of making the tide come in and go out. Although this surely is an effect of the moon, it simply does not warrant an ascription of function. We must have a way to block such ascriptions, and wedding the ascription of functions to goal-directed systems provides the means to do so. (Adams 1979, 496)

In contrast to the moon example, we do say that capillary constriction in the extremities has the function of restricting blood flow because such constriction occurs within a goal-directed system, a system directed to the goal of temperature regulation. Thus, adding the requirement that functions occur only in a goal-directed setting effects the desired difference. "The goal-state theory of function attributions holds that the central feature of functional relations is not only that they are means-end relations, but also that they are means-end relations within or conjoined with goal-directed systems" (Adams 1979, 496).

Some hold, he notes, that to be goal-directed requires mentalistic properties, such as having desires, beliefs, and intentions. Since functional descriptions are also applied to things with little or nothing by way of beliefs, desires, and intentions, as is the case of lower animals and plants, such a position involves metaphorically extending those concepts. This, Adams believes, makes the criterion for function talk unacceptably subjective. "Consequently, on *this* kind of goal-state theory of function attributions, whether a structure has a function will also be a purely subjective matter, if that structure occurs in a non-mental system" (Adams 1979, 497). However, whereas Nagel understood a self-regulatory system in terms of a state description involving disturbing and compensating variables, Adams, reviving the view of Rosenblueth, Wiener, and Bigelow, sees it as a cybernetic system. "I shall contend that a goal-directed system—whether mental or non-mental—is a cybernetic system" (Adams 1979, 505). The feedback signal is used to reduce deviance of the output from the goal state. "The system must also be able to process information about its present state . . . and it must be able to compare that information with its goal-state" (Adams 1979, 505).

Apparently because the goal-state is a future state and the future cannot cause the past, Adams, in a significant and promising advance over Rosenblueth, Wiener, and Bigelow, includes in the feedback theory provision for representation.

> Hence to be truly directed by a goal, and not merely uncontrolledly approaching an end-state, a system must have both an internal representation of a goal-state and a feedback system which conveys information about its present state, and its output values, back into the system as inputs. (Adams 1979, 505)

This resolves a major problem Taylor had discovered in the earlier version of feedback theory. Rosenblueth, Wiener, and Bigelow had said, "If a goal is to be attained, some signals from the goal are necessary at some time to direct the behavior" (1943, 19). In response, Taylor observed that this would make the analysis incapable of accommodating such common behavior as groping in the dark for matches when none were there or looking in the refrigerator for an apple when there was none. On Adams' account, the error signals come not from the nonexistent matches or apple, but from representations of them, and they exist even when the matches and apple, the goal-objects, do not.

Summarizing, a goal-directed system is a system that has

(1) an internal representation of the goal-state;
(2) a feedback system by which information about the system's state variables and its output values are fed back into the system as input values;
(3) a *causal dependence* between the information which is fed back into the system and the system's performance of successive operations which minimize the difference between the present state of the system and its goal-state. (Adams 1979, 506)

The representation-generated feedback must be operational, must actually affect output. Emphasizing that the concept of a feedback system defines what it is to be a goal-directed system, he adds, "And, it is the information feedback system plus an internal representation of a goal-state which are the central features of a goal-directed system" (1979, 507).

Using this concept of a goal-directed system, Adams summarizes his analysis of functions:

> A structure x has a function y *just in case*:
> (1) x does y in system S;
> (2) y causally contributes towards S's outputting O (*through the causal feedback mechanism*);
> (3) O is (or itself contributes toward) a goal-state of S. (1979, 508)

He refers to a thermostatically controlled heating system to illustrate how his analysis works, with the position of the dial representing the goal-state.

Some things have functions that are not themselves feedback systems, such as clips on ink pens or rocks used as paperweights. To provide for such things to have functions, Adams adds that the item need not actually be a part of a goal-directed system but may merely be conjoined with one.

> It is true that we appear to attribute functions to structures that occur in systems which are not goal-directed. To account for this fact I will adopt the view that it is only in conjunction with a goal-directed system that it is correct to ascribe functions to non-directed systems, e.g., to non-goal-directed artifacts. (1979, 508)

A clip on a pen has the function of affixing the pen to pockets because the clip is conjoined or associated with a goal-directed system. A certain rock has the function of holding down papers because it is associated with a goal-directed system having that goal. "What I wish to capture is the intuition that, e.g., although a rock may come to hold papers in place on a desk, it is only in virtue of being conjoined with a goal-directed system—Smith—that the rock comes to acquire this function" (Adams 1979, 509). Mere proximity is not enough, however. "Further, even if the rock comes to be associated with Smith, it must instantiate the appropriate means-end relation *through the feedback loop* to acquire a function." The conjoining creates a new and more encompassing system, S, that is, Smith and the rock together. "In any event, the system consisting of Smith plus the rock must be a goal-directed system—which Smith is. The rock becomes an artifactual extension of Smith" (1979, 509).

There are several difficulties with Adams' analysis. Beginning with the last first, moving from the original claim that for x to have the function y, x and y must themselves be a part of a goal-directed system to the later claim that it is sufficient that x and y are associated or conjoined with a goal-directed system constitutes such a major change that the formal analysis should have been reformulated. If one reads the analysis as including the later modification, that analysis must refer to rather strange and unusual systems. Even in Adams' account, it is Smith who has the goal of using a rock to hold his papers. Yet, "S" does not stand for Smith in the revised analysis, but Smith and the rock together. While it is possible to carve out a system such as Smith-rock, it is not that system to which we ascribe the goal of holding down papers. No one even ventures to guess what such an unusual system might do or what goals it might have. We have no name for such a system. We can only refer to it by describing its conjoined components. It reminds one of Nagel's conjoining field or environmental elements with the behaving organism to form a new composite system. Adams' composites, as Nagel's, violate our sense of natural kinds, which, after all, form the subject matter of science.

The awkwardness of such systems becomes even more evident when one imagines trying to explain the behavior of someone using several items successively, as a carpenter using first a hammer, then a saw, and finally a square. The explanation of the functions of these common tools requires a different system for each tool, now Smith-hammer, a moment later Smith-saw, etc. The linguistic contortions increase when one considers how the analysis would handle the functions of his shoes, belt, and shirt, all of which are used simultaneously. Because explanations that appeal to such evanescent and unusual systems are not credible, the analysis does not give a satisfactory explanation of the language of functions in the extensive realm of nonfeedback human artifacts.

The analysis also has difficulty handling functions of parts of organisms that, although parts of feedback systems, do not do what they do because they are parts of feedback systems, such as fur having the function of maintaining warmth, bones having the function of supporting the body, joints having the function of allowing movement of limbs, incisors having the function of cutting, molars of grinding, and so on. A bear's fur, unlike peripheral capillary size, does not have its function in virtue of being part of a feedback system. It is part of a system having feedback but is not part of the feedback mechanism. The fur does what it does in an uncontrolled manner, violating the second line of the analysis, which says that "y causally contributes to S's outputting O (*through the causal feedback mechanism*)." Expressing the matter in terms of goals, although the fur is conjoined to a goal-directed system, there is no connection between the goals of that system and the item having the function, no connection between the goals that bears have, even their goal of keeping warm, and their having fur. In this, the example is unlike that of the pen clip and rock, where the goal of the goal-directed system, the person with the goals of having his pen available and papers secure against drafts, is connected to the pen having a clip and the rock being on the desk.

On the other hand, being included in the feedback loop is not sufficient to give an item a function. A bush, let us say, conceals a rabbit, and is, therefore, associated with a goal-directed system, the rabbit, and in a manner involving feedback. Therefore, according to the revised analysis, it becomes correct to say that it is the function of the bush to conceal the rabbit. It, then, also becomes a function of a branch to support a sparrow and of an updraft to elevate an eagle, and, since plants utilize water in ways involving feedback, a function of rain to water the garden and, because functions need not be beneficial, drought to shrivel it. The revision makes the analysis too indiscriminating.

Adams begins his examination with the observation that "the essential feature of a functional relationship is that of a *means-end* relation" (1979, 494) and later says, "I am arguing that the primary feature of functional relations is their being means-ends relations" (1979, 510). It is surprising, then, that nothing about a means-end relation appears in his formal analysis, for one would think the analysis would include what was essential. Perhaps his explanation would be that means-end relations are purely causal and reference to causal relations does appear in his analysis. However, if, as he says, the cause-and-effect relation of heated and expanding gas is also a means-end relation, means-end relations would be ubiquitous even in inanimate nature and the concepts would be of no use, for ordinary causal relations would do as well. This makes it puzzling why he declares the means-end relation to be the essence of the functional relation. The importance initially placed on means-end relations is out of place with the insignificant role it plays in his final analysis.

Adams believes fortuitous effects are blocked by his means-end goal-directed system. However, since he sees means-end systems as purely causal and goal-directed systems as feedback systems, hence, also purely causal, he, in effect, says that fortuitous effects are blocked by certain kinds of causal systems or patterns. Put in these terms, his thesis is implausible. Suppose the wires from a thermostat to the furnace become corroded but that a book falls from the shelf above and lodges on top of the wires, its weight sufficient to maintain electrical contact between the corroded wires and the terminals. The book causally contributes to the thermostat's maintaining a certain temperature, and the thermostat is a feedback system, hence a goal-directed system. According to the terms of the analysis, the book, therefore, has the function of maintaining electrical contact at the terminals. Yet, its being there is a fortuitous effect.

To say x is a means to the end y is not at all the same as to say that x causes y. To say x is a means to the end y is to say that x is a way of reaching goal y. Means-end talk is goal talk. A rock holding up a corner of a cabin is a means of keeping it off the ground and preventing rot, but a rock holding up a tree blown over in a storm is not; yet both were caused. The difference is that keeping the porch off the ground and preventing rot is a goal or end but holding the tree off the ground is not. Apart from some context in which maintaining temperature is the goal or aim, a gas's expanding is not a means of maintaining constant temperature. Adams is correct in suggesting that there is some significance to the claim that functional relations show up only in the context of means-end relations, but that shows that functional relations are not merely causal relations and, since ends are goals, that functional relations show up only in the context of goals.

We routinely and with great agreement and consistency judge which objects are goal-directed without knowing their internal structures. Not only do people so judge, but animals do, as well; in fact, their survival depends on it. Since we routinely determine something to be goal-directed without knowing whether it uses feedback, the concept of a goal-directed system cannot require that the system be a feedback system.

In Adams' analysis, for the function of x to be y, the line "x does y in system S" requires that x actually do y, and the line "y causally contributes towards S's outputting O" requires that y actually contribute to O. He later repeats the position.

> If I build windshield wipers out of cardboard, it is not sufficient that I selected them for their ability to clean windshields, that they have that function. If they cannot enter into the means-ends nexus in virtue of which my car outputs O (cleans the windshield in the driving rain), then that is not their function. They do not actually have that function since they cannot deliver. (Adams 1979, 513)

Since his analysis requires success, it does not provide for an item to have function y if it does not actually do y, thereby leaving out the immense class of faulty artifacts—knobs that do not turn, lamps that do not light, and motors that do not start. It is distressing how long-lived and persistent in the literature this error is. An item could not be defective if function disappeared with performance. Customers are upset with defective merchandise only because function continues without performance. Without the concept of function, a knife that cannot cut would be of no more interest than an oak leaf that could not cut. The concept of malfunction is based on the assumption that something can genuinely fail to perform its function, which it could not do if its function were absent whenever y was not produced. Functions vanishing with failure would make things awkward for items that worked intermittently. An automobile engine running and stopping ten times in ten minutes would lose its function and regain it ten times in ten minutes. Malfunctions may be of such brief duration, but not functions. Bringing into the analysis reference to tendencies and probabilities would not help, because some items have as functions things they cannot do.[4]

Adams opposes an intentional analysis of functions. "It seems that one's intentions (i.e., the fact that one selects x for y) are neither necessary nor sufficient for x's having the function y" (Adams 1979, 512). Malfunctions are difficult to accommodate. Denying that function continues when performance falters has the consequence that it is not necessary to deal with malfunctions. The fact that things do malfunction, genuinely fail to perform their functions rather than merely lack them, should make one question whether intentionality is dispensable.

Although he rejects the intentionality of functions, he accepts their intensionality. The function of x may be y and y may be identical to y', yet it not be the case that the function of x is y'. Using his example, consider a computer programmed for letter recognition, with a detector for each letter. The function of a certain detector is, let us say, to detect "Q." Even though "Q" is the seventeenth letter in the alphabet, it is not the function of the detector to detect the seventeenth letter. His analysis, he says, preserves the intensionality because it is the shape of the letter that causes it to be detected, not its numerical position in the alphabet:

> Although it is undeniably true that if the device detects a Q it detects the seventeenth letter of the alphabet, it is not *in virtue of* a Q's being the seventeenth letter of the alphabet that the device detects it. Rather, it is in virtue of a Q's having the physical configuration which it does that the device detects it. (Adams 1979, 516)

In order for an example to show that his analysis preserves the intensionality of functions, the example must fit his analysis. The function claim said simply that the function of the detector was to detect "Q." It said nothing about how that detection was to be achieved. The example, following the language of his analysis, should say:

(1) the detector detects "Q" in system S;
(2) detecting "Q" causally contributes towards S's outputting O (through the feedback mechanism);
(3) O is (or itself contributes toward) a goal-state of S.

The relevant part is line (1). The term "detect" threatens to compromise the analysis because it is not clearly naturalistic. The line must here be given a nonmentalistic reading, such as "The detector causes the detection of 'Q' in system S." Using the same form for the seventeenth letter example, line (1) would be:

(1) the detector detects the seventeenth letter in system S.

It is also to be interpreted causally, such as "The detector causes the detection of the seventeenth letter in system S." However, that statement is true, not false. Indeed, Adams acknowledges this when he says, "It is undeniably true that if the device detects a Q it detects the seventeenth letter of the alphabet." The fact that when the causal claim about detecting "Q" is true, the causal claim about detecting the seventeenth letter is also true means that Adams' example fails to show causation to be intensional. Since his analysis treats functional relations as purely causal and functional relations are intensional, this means that his analysis fails to preserve the intensionality of functions, and suggests that any purely causal analysis will fail.

Adams says, "Let us assume that we build this device so that it detects Qs by analyzing their physical configurations" (1979, 516). Of course, a causal event can always be further analyzed. If the detector detects the letter "Q," it must do so in virtue of some property or other. However, since a function statement does not specify in virtue of what property the function is to be realized, does not specify its own mechanism, it is incorrect to add that information in the analysans.

Further, it is incorrect to claim that if the detector detects the letter "Q," it must do so in virtue of shape. It could detect that letter by means of any of a large number of properties, such as being the tenth letter from the end of the alphabet, being the first letter of the alphabet after the letter "P," being the letter usually followed by the letter "U," being the first letter in the word "quark," or even in virtue of being red, if all the occurences of the letter "Q" in the sample were red and nothing else was.

One must also remember that the shape of the letter "Q" varies greatly from font to font, printing to cursive, handwriting to handwriting, century to century. In fact, since the detector is described as being a computer, it presumably would be a digital electronic computer, in which case the detection would not be by shape at all but by a certain sequence of voltage pulses or square waves. According to the ASCII code, these would be isomorphic to the binary 0101 0001, with "0s" matched with the voltage lows and "1s" with the highs. There is nothing here about shape.

Even if adding the information in the causal analysans were not objectionable, doing so would not help. Suppose the analysis specified that x causally, utilizing shape, produces y. If y were identified in the analysis as the letter "Q" and were not identified as a Q-shape, it would still be the case that the output of x would be the seventeenth letter of the alphabet. Specifying how the causation occurred would not alter the extensionality of the effect. Although Abraham Lincoln died by a bullet rather than a stroke, that did not alter the fact that when Lincoln was killed, the president of the United States was killed. The only way outputting the seventeenth letter could be blocked would be to specify the shape of the letter "Q" as the output, not the letter "Q." That would, however, be useless since the shape of "Q" is not identical to the seventeenth letter of the alphabet, though it might be identical to the shape of the marks on a butterfly wing or the shape of a segment of encircling ivy. If the output were the letter "Q," itself, then the output would also be the seventeenth letter of the alphabet, regardless of the components of the causal chain.

Adams' example would not have appeared plausible if what was detected had been something having the same shape in two languages, as the English "o" and Greek omicron or English "X" and the Greek chi. The computer's function is, let us suppose, to detect the English "X," and it does so by pattern recognition. However, in outputting the English "X," it also outputs the Greek chi, but its function is not to detect chi. Therefore, causation, even when operating in virtue of pattern recognition, does not underwrite function. If one objects that the detector did not output chi but only the English "X" because only that was intended, then, of course, the analysis would no longer be in terms of causation, but in terms of intentionality.

Causation is not, in general, intensional. If x causes y and y is identical to z, then x also causes z. If hitting an iceberg caused the sinking of the *Titanic* and the *Titanic* is identical to the world's most luxurious ship, then hitting an iceberg caused the sinking of the world's most luxurious ship. Restricting the causation to a certain kind does not alter this fact.

As noted above, Adams requires that the system be able to represent the goal-state, but in explaining, says, "When we set the thermostat at 65°, the position of the dial becomes an internal representation of the goal-state of 65° room temperature" (1979, 507). However, Ehring notes that if the dial were damaged, the thermostat might actually maintain the temperature at 80 degrees. "In that event we cannot without further argument take the goal-state to be 65 degrees. Adams' suggestion that we can determine the goal-state directly from the dial setting presupposes that the system is in 'good working order'" (Ehring 1984, 219). Determining what constitutes good working order would, he adds, lead to circularity. It might seem that one could avoid the circularity by appealing to the state the system tends to reach, but that response would be vulnerable to examples involving systematic error, and circularity would soon reappear in the explanation of what constitutes systematic error. Determining what is good working order on the basis of what is normal or common would not suffice with examples in which poor working order were pervasive, as in the case of widespread illness, or, with artifacts, poor quality control. Determining what is good working order in terms of natural selection would not apply to artifacts employing feedback, including servomechanisms (Ehring 1984, 219-20). To his reasonable reservations might well be added the likelihood that representation carries with it all the complexities and problems of teleology and that understanding one will likely carry with it understanding the other, as well.

Adams' analysis, though having its problems, is certainly an interesting one, especially in view of the inclusion of provision for representation. Representation seems essential for functions and teleology generally, not because a feedback analysis needs something to emit reference signals, but because of the need to account for the future orientation of teleological language without requiring reverse causation. Because functions seem to require representation, it appears unlikely that current attempts in philosophy of language to base a theory of representation on biological functions will succeed in avoiding circularity.

Faber

Roger Faber, a physicist, offers an informed and thoughtful account and defense of the feedback analysis of teleology. He examines the structure of a negative feedback system in far greater detail than have other analysts, identifying and labeling its many properties and relations. He is motivated in part to respond to Wimsatt's challenge that the concept of feedback is not well understood and to improve on the work of Beckner.[5] Summarizing to a bare minimum, there is a controlled system, S, a sensing-controlling subsystem, R, and a property maintained within a certain range. When disturbing influences enter S from the environment, the sensing subsystem detects deviations in the controlled property and, by means of a negative feedback causal loop, the controller modifies the system S, with the result that the property is maintained within the protected range. In a major departure from previous accounts, Faber adds the condition that the sensing and controlling systems be each separately physically detachable from the controlled system.

Goal-directed behavior and functions are both analyzed in terms of feedback systems. The goal of a feedback system is the state it tends to maintain, and it must be a possible state. "If a mechanism does seek a goal of its own, the sought-for condition must be a state of affairs that the action of the mechanism tends to achieve under some possible circumstances" (Faber 1986, 77). The goal need not be the human goal for that machine. The goal of a thermostatic heating system is to keep the bimetal strip at a certain temperature, whereas the goal of the householder, another feedback system, is to keep the house at a certain temperature (Faber 1986, 77). Faber agrees with Woodfield that a goal-directed system must be able to represent the goal. However, he adds that it is sufficient if that representation is representation to the system, itself; it need not be representation to the observer.

Functions are related to feedback in that the parts and processes of the feedback system that contribute to achieving the goal of that system have functions.

> When we ascribe a function to a portion of an organism or a machine, this is what we do or ought to do: We call attention to the fact that the containing system tends to maintain some variable property within a narrower range than would otherwise obtain, and it does so by means of a peculiar sort of pattern of causal connections, namely, negative feedback. (Faber 1986, 89)

Thus, functions also receive a feedback analysis, though a slightly different one from that of goal-directed behavior.

> Only those things and processes that contribute to the normal operation of the loop (whose removal would impair or halt it) have functions with

respect to that goal. The function of a member of a feedback loop is simply what the part contributes (not by happenstance but in the regular causal chain) toward the goal embodied in the loop. (Faber 1986, 89)

There are functions only where there are goals, and goals are what devices with a negative feedback loop tend to maintain. His division of labor, however, is not sustained, and what parts do is treated both as functions and as goal-directed behavior.

Faber claims that the feedback concept is sufficient to explain functions and goal-directed behavior, but, at least usually, does not claim that feedback is necessary (1984, 49, 117).[6] The ultimate goal of organisms, he believes, is individual survival, and a feedback analysis very effectively accommodates functions and goal-directed behavior of those subjects with that goal. However, since what promotes reproduction, such as feeding and protecting the young, increases risk to the parent, feedback does not, as he sees it, account for the functions involved in reproduction. "It pronounces them to be maladaptive, dysfunctional" (Faber 1986, 103). Although he criticizes at length an analysis of function based on selection, he does, especially in his second work, *Clockwork Garden*, accord selection the primary role in grounding functions associated with reproduction. "Yet the fact remains that selectionism makes sense of reproductive functionality, and cyberneticism does not" (Faber 1986, 110). His considered view, then, is that negative feedback covers most functions and that negative feedback plus selection cover all.

> Are there other types of organization that would also justify function ascriptions? . . . Possibly so, but I submit that the two cybernetic patterns so far discussed are general enough to cover the scientifically interesting cases. Supplemented by the concept of selection . . . they will prove adequate to make sense of all the usages—even the weakly metaphorical ones—of functional language for natural systems. (Faber 1986, 92)

Nevertheless, it is unclear just how comprehensive he intends a feedback analysis to be. On one hand he says, "The theme of this essay is the mechanism of goal-orientation in non-conscious organisms and in machines" (1984, 44),[7] suggesting that he is not talking about functions involving human consciousness and that of the higher animals. However, he also says, "But with cybernetical concepts we have constructed a thoroughly mechanistic, hence reductive, explication that accommodates all teleology, from human purposes to means-end relations in the simplest living things" (1986, 120). Furthermore, he freely uses examples from human physiology and behavior and from biology in general.

Faber presents his analysis as reductionistic, replacing the concepts of goal-directed behavior and functions with that of physical causation. "I have claimed that my explication of feedback is reductionistic because it is done in the language of ordinary physical analysis" (Faber 1984, 111). He characterizes his analysis as a causal account in which the causation exhibits a certain pattern. "Accordingly, I suggest that to give a functional explanation of a system is to recommend a selective way of looking at it, a way in which some causal connections are emphasized and others ignored" (1984, 110). The goal of an organism or a mechanism can be determined entirely by its structure and activities, with no reference to the intentions of either the designer or the user (Faber 1986, 76). "I have required that any goal we ascribe to a mechanism must be discoverable just from an inspection of its inner workings" (Faber 1986, 85). He reiterates, "What matters for reductive atomism is that a thing's being a goal-seeking system be entirely an objective property of the system itself, with no trace of human intentionality being impressed upon it from outside" (1986, 96-97).

The separation of his feedback analysis from anything involving intention and conscious states is, however, compromised when he appeals to context in identifying functions. "I shall argue in this section that the context helps determine functionality in several ways" (Faber 1984, 101). Context is not limited to physical surroundings, but may include expectations and intentions. He specifically allows that context may involve use. A circuit that regulates voltage also regulates current. Is it a voltage regulator or a current regulator? "Now it may be the case that, owing to the details of construction of the device, it would make a better current regulator than a voltage regulator. Nevertheless, if it is used as a voltage regulator, a voltage regulator it is" (Faber 1984, 102). Here, use determines function. The concept of use, however, is an intentional concept. A hammer is used to pound in a nail only if one intends to pound in a nail with it, not if one is trying to break the hammer, even if the nail ends up embedded and the physical description is the same. A car is used to get to the store only if one intends to get to the store with it, not if one is testing its brakes, even if one happens to end up at the store and the physical description is the same. Similarly, a circuit is used as a voltage regulator only if there is the intention to do so. The intention need not, of course, be conscious; habitual uses often are not.

In addition to the context's affecting the determination of function by means of use, Faber also speaks of context in terms of the designer. "Some context of electronic design and use is required to justify calling this circuit by its functional designation" (1984, 102). "In the most general case, when we ask what a part in some artifact is there *for*, the answer can be obtained only from its designer" (1986, 97). The point of bringing in the designer, however, would hardly be to examine the shape, properties, and arrangement of parts, for that, presumably, had already been done without determining function. Surely the designer is consulted regarding his intention in producing the design.

As noted earlier, he says that the goal of an organism or machine can be determined from observing its inner workings. This assumes that observation alone is capable of revealing defects. Perhaps the wire to a terminal of the thermostat is loose. How does one know that being loose is a defect? One needs to know how things should be, but observation reveals only how things are.

One could, of course, accept the fact that inspection alone cannot reveal defects, since defects can be inferred only by comparing with how things should be, while still insisting that observation can reveal goals; however, then one must also hold that the devices can never fail to reach their goals. That is, if goals are determined by structure and are detectable by observation, and if one cannot observe defects, one could not read off a goal that had not been achieved. Stated another way, failure to achieve a goal is ruled out logically because that would require a goal that was determined by structure (to meet the requirement of observability) but was not determined by structure (since the structure did not achieve the goal). Thus, if failure is possible, which Faber assumes throughout and specifically allows by his use of "tends," then defects must be observable; but if so, there must be independent knowledge of the goal or function, apart from and not reducible to observing the inner working of the machine. It is difficult to locate any plausible source for this knowledge outside of the probable intentions of the designer or user.

When Faber allows for context and for observing defects, intentionality is introduced by the back door. Indeed, he momentarily acknowledges a place for intentionality when he says, regarding a small hammer hanging by a fire alarm, "The function of the hammer is determined by what the designer and users of fire alarms intend for it" (1986, 97).

Nevertheless, Faber argues vigorously against intention having any role in determining function. To support his case, he imagines a feedback system produced accidentally. In one example, a voltage regulator is assembled by random motions. "But I claim that, given this context, even a circuit produced by chance would qualify" (Faber 1984, 102). If there were merely a general context of electronic design, it would, he says, be correctly described as a voltage regulator even though there were no intention to produce one or use something as one. Therefore, intention is not sufficient for function.

The terms in which the example is described, however, beg the question. Faber's description begins with, "An isolated voltage regulator circuit, formed by a random collocation of atoms" (1984, 102), guaranteeing that there is a voltage regulator. If one has a voltage regulator, then, of course, regardless what else might be the case, one has a voltage regulator. To avoid such circularity, the conjecture must begin in such a way that it is not known what the object is. This might be accomplished by describing the object in terms of materials and shape.

A second misleading feature about the thought experiment is that the item in question is so complex that we have difficulty imagining its being produced by random events. We are told to imagine a voltage regulator circuit produced at random, and we immediately think of a coil of insulated copper wire, knowing full well that copper does not occur in nature as insulated wire. We try to imagine that this substance that, without processing, just happens to be not copper ore but pure copper, has accidentally been drawn into wire, and, as a result of a long series of coincidences, somehow ends up with a cover of electrical insulation, and lies evenly and neatly coiled around something that by pure happenstance is soft iron, which, by the vagaries of tide and tempest, has been shaped into a cylinder of dimensions we associate with precise machining. Of course, we have difficulty performing this remarkable feat and end up fudging the task, not clearly imagining things as randomly produced when we know making them requires great skill and effort, believing we are imagining them when we are really only running over the words. It is good to remind ourselves of the occasional student who insists he can imagine a square circle.

In another example, Faber has a sculptor putting together electronic components in various ways, accidentally producing an ensemble that can regulate voltage. He infers that its function is to regulate voltage because it is able to do so, and, since it was produced unintentionally, he concludes that function does not depend on intention. "I think we must be willing to say both that it is a voltage regulator and that the function of the little spring-loaded coil is to adjust the setting of the variable resistor" (Faber 1984, 87). A questionable feature of this carefully crafted example is that the components—the coil, the switch, the wires, the resistor, and the bimetal spring—are all products of considerable design. The accidental aspect is severely restricted, the intentional quite prominent. In spite of that, it is unlikely we would judge the result a voltage regulator unless we fudged again and did not genuinely and rigorously comprehend the collection as accidental. He needs a fairly complex object, such as a voltage regulator, thermostat, or clock, but one formed from natural forces and processes, such as gravity, heating, and erosion, to make his case. These demands are hard to fill because we immediately take such well-ordered complexity to have been produced intentionally on the grounds that accidental production is unlikely beyond belief.

One could extend Faber's examples of accidental construction and reduce the accidental element to a minimum by imagining a futuristic factory that manufactures various articles, for example, thermostats and clocks, by pressing either of two buttons. The wrong button could be pushed inadvertently, thereby producing a clock accidentally. With such a small accidental component, we would no doubt continue to regard that clock as having the function of keeping time. However, this would do little to threaten the role of intention in determining functions, for the clock produced by mistake would still be primarily the product of design.

Instead of being about voltage regulators, the thought experiment should have been about something simple enough so its production by chance would have been credible. Since Faber extends his analysis to things that are not themselves feedback structures, such an example should be acceptable. Consider a piece of quartz, which, due to weathering and erosion, is now roughly six inches long and two inches wide, with one sharp, jagged edge. Even given the context that it is found in a stratum containing remains of a culture that made stone knives, anthropologists would consider it an error to judge the item a knife, given that it is known, as assumed in this example, to have been the product of weathering and erosion. The reason they would surely give would be that the shape had been produced accidentally, which is to say, not intentionally. It might still be judged a knife, however, were it found in conjunction with cooking utensils, but then it would have been because of presumed use.

Another example Faber uses to show that intention is not sufficient to determine function is the highly unlikely one of an accidentally produced thermostat.

If some inept fabricator assembled a thermostat by inadvertence,
intending to make some other sort of contrivance, all his protestations
to the contrary would not overrule our claim, based on an analysis of
how the device operates in fact, that the function of the relay switch is
to turn the furnace on and off and the function of the bimetallic strip is
to sense the temperature of the room around it. (1986, 97-98)

The story illegitimately guarantees that a thermostat can be produced
without intention by identifying it at the start as a thermostat.
Furthermore, thermostats are defined in part in terms of how they are
used, and use is an intentional concept. Again, the example is of
something of such complexity that, entropy being what it is, imagining it
produced by ineptitude and fumbling invites confusing words with deeds.
A better example would have been something simple and crude enough
for production by fumbling to have been plausible.

Faber also argues that knowledge of intention is not needed because
goals can be identified by the machinery. "Can an organism's or a
mechanism's goals be identified by reference solely to the structure and
activities of the thing itself, apart from the intentions of its employer or
designer? . . . I respond with qualified affirmatives to these questions"
(1986, 76). Again, "An object or a dynamical process may *aim* at a
goal—that is, it may be a full-fledged teleological system" (1986, 82), for
example, "an incandescent lamp placed too close to this feedback device,
or an ice cube balanced on top of it, will soon reveal that the machinery
itself, if it aims at anything at all, aims at a temperature of 20° only for
its own bimetallic strip" (1986, 77).

If the goal were created simply by a feedback device causing some
state to be maintained within a given range in the face of disturbing
influences, the goal in the example would not be limited to the
temperature of the bimetallic spring. The shape of the spring, the angle
of contacts from the vertical, the distance of the upper contact and the
nearest point of the housing, the temperature of the housing, and on and
on, all are maintained in certain ranges in the face of disturbing
influences because of a feedback system. Adding a lamp or ice cube
would merely make the house temperature settle down to a different
equilibrium. There would still be an indefinitely large number of goals
pertaining to the temperature at various locations. If a goal were created
merely by a feedback device causing a state to be maintained within a
certain range in the face of disturbing influences, keeping the air one inch
from the ceiling in one temperature range would be a goal, keeping the
temperature of the inside of the north storm window in another range

would become another goal, keeping the length of a metal curtain rod within a certain range another, and on and on. Even the amount of heat radiated from the house would become a goal, something the householder does not even desire. Thus, even if it were established that all goal-directed systems exhibit feedback, the feedback analysis is simply not sufficiently discriminating for analyzing goal-directed behavior.

A major difficulty with a feedback analysis of functions arises from the fact that just about anything can have a function, including things that are not feedback systems, things such as hammers and doorknobs, fur and hooves. Faber, like Adams, is aware of this, and addresses the problem in a similar manner by saying that, in such cases, the item in question has a function because it is aimed by or directed by a feedback system. Responding to my claim that a hammer's function cannot be explained in terms of a self-regulatory system (Nissen 1980-81, 134), he writes,

> We must distinguish two senses of the term "goal-oriented." An object or dynamical process may *aim* at a goal—that is, it may be a full-fledged teleological system—or, like the hammer, it may *be aimed* by a system of the former sort. Both the system that actively aims and the one that is aimed may qualify as goal-directed by reference to a cybernetic analysis of the former. (Faber 1986, 82)

His replies to the question about how a hammer can have a function by explaining how it can be goal-directed, thus merging the two concepts. The fact that a hammer can have a function without being a feedback system is claimed not a counterexample to the feedback analysis of functions because a hammer is aimed or directed by a feedback system. The hammer has a function derivatively.

Earlier he said that a goal is what a feedback system, utilizing its feedback mechanism, tends to maintain, and a function is what a part or a process of a feedback system does in contributing to that goal. A hammer, not being a part of a feedback system at all, is neither one nor the other. In response to my calling attention to the large class of nonfeedback artifacts represented by the hammer, he radically extends his theory so that it is no longer required that the item functions within a feedback system; it is enough that it is directed by one. Since humans are safely feedback systems, this takes care of all human artifacts, but now the role of feedback has been compromised. In what is presented as merely the drawing out of consequences of his original theory, Faber produces, as did Adams, what is really a second and much expanded theory of functions. It is a feedback theory at second hand, since the item having the function in the new sense plays no role within the feedback system.

It must be acknowledged that we do give the function of a hammer a derivative status, present because of the people who designed it, made it, or use it. However, we, if pressed, explain that function by linking it to the intention of those persons.

The functions of many parts and processes of an organism, such as insulin and thyroid secretion, are covered by the original theory because they operate as working parts of feedback systems. However, there are many parts of organisms that, though contributing to survival, do not utilize feedback and are not directed by a feedback system, hence fall outside both the original feedback function theory and the revised theory. The parts of the skeleton, the teeth, and the scaly, feathered, or hairy epidermis, as well as external coloration, are of this kind. The function of the aperture in a vertebra is to provide room for and protect the spinal cord. This function is not covered by the original theory because, although part of a feedback system, it does not contribute to survival by means of feedback, and it is not covered by the revised theory because it is not aimed or directed by a feedback system. The functions of camouflage and mimic colorations are similarly exceptions.

Parallel to the second theory of functions is a second theory of goal-directed behavior: "We must distinguish two senses of the term 'goal-oriented'" (quoted above) and "We must distinguish between the goal-directedness of a whole system and the goal-directedness of some of the behavior and internal processes of the system" (Faber 1986, 80). That he offers two theories receives added support from his response to Woodfield's criticism of Rosenblueth et al. that some behavior is goal-directed without having feedback. Woodfield uses the example of a cuttlefish's striking at prey with its tentacles. The action occurs too rapidly for feedback to play a role, yet is surely goal-directed. Faber allows that the action does not involve feedback but maintains that it is still goal-directed, and even adds the examples of a person's reaching for a coffee cup without looking and a relay's closing in response to an impulse from the thermostat.

He explains by alluding to the original feedback theory, "According to the cybernetic theory, a system is oriented toward a goal by virtue of its organization in the feedback pattern," and continues, referring to what is in fact a second feedback theory, "a sample of behavior or an internal process may be said to be directed toward that goal if it contributes to its achievement" (Faber 1986, 80). The cuttlefish's striking, the person's reaching for the coffee cup without looking, and the relay's closing are samples of behavior of a feedback system, and that is what gives them goal-direction, even though no feedback influences those behaviors. Generalizing, behavior not itself guided by feedback but that is performed by a feedback system and contributes to the goal of that system is also goal-directed. This is to be contrasted with behavior guided by feedback, such as reaching for a coffee cup while looking, which is covered by the original theory. It is remarkable that two actions so much alike as reaching for a cup while looking and reaching for it while not looking should require different theories.

Ehring criticizes the feedback theory in a way similar to that of Woodfield, saying that behavior is sometimes obviously directed to a goal even though no feedback is possible.

> Suppose that John has lost all kinesthetic feeling in his arm. Without being able to see or feel the position of his arm and hand John is asked to wiggle his hand in order to activate an unseen mechanism which reacts to movements. John makes a conscious effort and unbeknownst to him succeeds in wiggling his hand and thereby activating the device. . . . His action is goal-directed despite the fact that John has no access to the effects of his activity. (Ehring 1984[a], 218)

With loss of feeling in his arm, there is no feedback, yet the action is goal-directed. Hence, goal-directed behavior does not depend upon feedback.

Ehring also considers an example not based on loss of sensory ability. Instead of a goal-directed action happening so rapidly that adjustment via feedback is impossible, consider action in a social setting that makes adjustment impossible. "Suppose that Jones seeks to win a one-time lottery by buying a ticket. This action is of course goal-directed, i.e., he buys the ticket in order to win the lottery. This action will have one of two outcomes: winning or losing the lottery" (Ehring 1984[a], 218). No succeeding adjustments are possible. There can be no on-course corrections once under way. "Where the goal-state is some particular non-repeatable state and the means consist of setting in motion a causal process over which the system has no control (except for starting it), then the goal-directed activity will not be part of a negative feedback loop" (Ehring 1984[a], 218).

Presumably, Faber would reject both as counterexamples and declare the feedback analysis sustained because both wiggling the fingers and buying the lottery ticket were performed by feedback systems. However, this response would fall short of extending the feedback theory to these cases in any significant way because feedback plays no role in the behavior. Feedback is a present but nonparticipating feature.

It may be that Faber would not acknowledge any difference between parts of feedback systems that have functions, such as thermostat relay switches and hearts, and things that are not parts of feedback systems but still have functions, such as hammers. Perhaps, instead, Faber, like Adams, would regard a hammer also as a part of a feedback system by construing it as belonging to a larger feedback system that includes the carpenter. But this way lies madness. As remarked earlier, if a hammer had a function because it is part of a larger feedback system, then one would have to countenance such bizarre systems as the carpenter plus the hammer, the carpenter plus the saw, the carpenter plus the tape measure, and so on. This strained position implies that as the carpenter lays down the hammer and picks up the saw, one system ceases to exist and another comes into being. Such transient feedback systems would be utterly unlike the examples Faber used when introducing the subject of feedback, which were stable and enduring. As mentioned earlier, it is an indication that we do not, in fact, countenance such systems that such evanescent units have no names in our language.

As noted in the discussion of Adams, if its contributing to the goal of a feedback system were responsible for a hammer's having a function, then a branch's contributing to a goal of a bird, which is a feedback system, should be responsible for the branch's having a function. Yet, we do not attribute such a function to a branch. Similar examples can be constructed about the oxygen in the air, gravity, solar heating, water, etc. If perspiring has a function in virtue of its being a part of a feedback system and contributes to the goal of cooling, then the fact that water takes up heat when evaporating should have the same function, for that also contributes to the goal of cooling and does so within the same feedback system. For some reason, we say the hammer has a function but not the branch, oxygen, gravity, or evaporative cooling, and Faber's analysis does not say why. A theory based on intention does.

Faber seems aware of this oddity because he says that a function must be an extra or added feature of something: "A functional part must be a gratuitous addition to the system, superimposed upon mere physical necessity" (1986, 95). He explains the arbitrary feature by identifying it with the detachability of the controlling system from the controlled system. This is incorrect, for all sorts of things besides the controller have functions, for example, the ignition device of the furnace, the damper, and the flue, and hence must have this added or arbitrary status. He even says, "The part must be present in the system and participate in its activity *as if* by design"(1986, 95), offering the surprising line of reasoning that the appearance of design is evidence of its absence.

The original theory of functions ascribes functions only to the parts of a feedback system, not to the whole feedback system. What the entire feedback system contributes to, on this view, is its goal. Our practice says otherwise, for we frequently speak of a feedback system as a whole having a function. In fact, we are more likely to say that maintaining the temperature in a certain range is the function of a thermostatic heating system than its goal. The same is true of some but not all organic feedback systems. It is true that to the animal, which is a feedback system, we attribute goals. However, we are far more likely to speak of the function of the digestive, respiratory, or circulatory systems than of their goals, and these systems are all feedback systems. The concept of feedback systems, suggestive as it is, simply does not provide a satisfactory instrument for dividing functions from goals.

Another problem arises when Faber conceptually links having a goal or a function to reaching a goal. He states this position in many places. In one, "The theory asserts that an organism or other system is oriented toward a goal state just if the parts are so arranged that under certain circumstances it would act in a way that tends to make the goal state occur or more nearly occur" (1986, 61). In another, "If a mechanism does seek a goal of its own, the sought-for condition must be a state of affairs that the action of the mechanism tends to achieve under some possible circumstances" (1986, 77). Again, "a sample of behavior or an internal process may be said to be directed to that goal if it contributes to its achievement" (1986, 80). And again, "When we ascribe a function to a portion of an organism or a machine, . . . we call attention to the fact that the containing system tends to maintain some variable property within a narrower range than would otherwise obtain" (Faber 1986, 89).

Since the theory requires either success or the likelihood of success for something to have a goal or a function, the theory carries the awkward implication that only healthy or nearly healthy organisms have parts possessing functions, that a severely damaged heart, lung, or kidney has no function. Similarly, his view implies that a thermostatic heating system has a goal (rather, a function) only in those heating systems that work or usually work. This is incorrect. We do not, as a matter of fact, cease to ascribe functions or goals to diseased organs or the broken or ineptly made artifacts. As noted several times, teleological attribution requires neither success nor probability of success.

Faber argues energetically against basing teleology on natural selection. Where other writers see selection, he sees only sorting.

> But nature is in no way like a kindly but nonpersonal Luther Burbank. Consider a swimming coach who selects her team by throwing the entire freshman class into the pool and signing up those who float. The coach selects, because she expects her charges to win a few swimming contests, but the pool only sorts. Nature does not select either; it does not look beyond the present scene of carnage and starvation. (Faber 1986, 116)

He speaks of natural selection as "Darwin's romantic metaphor" (1986, 116). It is not genuine selection because there is no criterion of selection: "Those genotypes which survive in the next generation are separated from those which do not. But what criterion does this separation mechanism serve?" (1984, 91). Feedback systems can select because they have a future orientation, but nature can only sort. "A selection process is teleological, not mere sorting, only if it involves at least two distinct things, a selector and a selectee" (1986, 115). He imagines a mutant earthworm that, instead of a heart, has cilia that caused internal fluids to circulate but less efficiently than the heart and are, therefore, maladaptive. It is clear, nevertheless, that the cilia have the function of circulating the fluids even though there was no history of natural selection (1984, 95). He sees as a difficulty for an analysis of functions in terms of selection the fact that we can determine functions of organs without knowing the organism's history.

Faber's vigorous criticism of selection theories, however, generates another problem for him. As noted earlier, he has concluded that goal-directed behavior and functions that relate to reproduction cannot be grounded on feedback. "In fact, the cybernetic theory of functionality treats reproductive structures even less hospitably than this example suggests: It pronounces them to be maladaptive, dysfunctional" (1986, 103). Since the structures and behavior needed for reproduction and rearing the young contribute to species survival but not agent or parent survival, the only way to extend the feedback analysis to them seems to be to construe the species as an immense scattered feedback system. Presumably Faber did not consider that an attractive solution. The only alternative he feels he can consider is an analysis based on natural selection; but that is the very theory he has so thoroughly criticized. So, although he rejects natural selection in the analysis of such goal-directed behavior as pursuing prey and for such functions as that of the heart and pancreas, he uses it for such activities as mating and feeding the young and for the function of such items as the bright plumage of the peacock. "Yet the fact remains that selectionism makes sense of reproductive functionality, and cyberneticism does not" (1986, 110).

He seeks to ameliorate the apparent inconsistency by saying that teleological language about reproduction is metaphorical and should be used within inverted commas. If the teleological language were metaphorical, there would be no need for basing it on either selection or feedback. Further, that solution has the awkward consequence that the term "function" in "The function of the ovary is to produce eggs" means something quite different from what it means in "The function of the heart is to circulate the blood," for one is a metaphor and one is not. Such a difference is just not credible. Adding to the impression that the analysis is in difficulty is the fact that Faber gives no indication in describing examples of reproductive functions that he is talking metaphorically. He talks of the function of the estrous cycle just as he talks of the function of binocular vision.

Any analysis that bases teleological language on recent discoveries must address the problem of how the new theory relates to traditional usage. Talk of functions and goal-directed behavior has been around a long time. Teleological language appears to be applied to the same things today as long ago. Usages that were controversial earlier are controversial still. This seems at odds with claiming that teleological ascription applies only to systems with feedback, since feedback organization has only recently been recognized. Faber dismisses the problem by alleging that the solution lies in the fact that both his recommended usage and traditional usage are bridged by the concept of survival (Faber 1986, 89-90). To the extent that analysis of functions is by prescription, then, of course, anything goes, and there is no way to judge one analysis right and another wrong, but a prescriptive analysis provides little reason for committing much time and effort to it. Where the teleological debate has been interesting, it has been in a setting in which the analyst works against a background of standard judgments about what is and what is not teleological. If one accepts traditional judgments, it is hard to see how a correct analysis of teleology can be in terms of features unknown to our predecessors. Even now, teleological judgments are typically, confidently, and accurately made by speakers quite ignorant of the subject's internal causal pathways. In arguing against basing teleology on selection, Faber writes, "But one can usually discover the function of an anatomical feature simply by examining how the machinery works here and now. Therefore, an evolutionary interpretation entails a conceptual break with biologists of the past, because they attributed functions without suspecting the evolutionary origins of living things" (1986, 111-112). Yet the feedback theory is vulnerable to a similar charge.

The problems about the concept of representation raised by Ehring in regard to Adams' analysis are also applicable to Faber's. Saying the representation of the goal-state is representation to the feedback system, if not meant metaphorically, raises a host of questions. What feature of the mechanism performs this representing? Surely, in the case of a thermostat, it is not the dial of a thermostat, for that suggests the mechanism can read. If not the dial, what? Apart from that problem, how is the goal-state picked out? Final states are not always goal-states, and understanding internal mechanism and reading blueprints, even thermostat dials, require intelligent interpreters. If all that is meant is that something occurs that is reminiscent or in some degree similar to representation, then there is no need for the rest of the analysis, for teleological language is immediately and without further question acceptable if used metaphorically. Appealing to metaphor whenever the theory is distressed is hardly satisfactory.

Nagel's self-regulation analysis proved vulnerable to my example about the temperature stability of an air mass subject to daily heating (Nissen 1980-81, 131). Faber believes that his feedback analysis, because it includes a requirement that the sensory and controlling subsystems be detachable, will block this counterexample. He illustrates his detachability requirement by diagrams of various kinds of closed vessels of heated water, the simplest of which is a covered saucepan (1986, 71). The selected protected property is maintaining a certain range of steam pressure. The sensory action is the lifting of the lid, and the controlling action is the same. Since they are the same, the sensory and controlling mechanisms cannot be separately detached, and the covered saucepan is, therefore, not a feedback system. In contrast to this, another closed heated vessel has a piston that moves with increasing steam pressure and is connected by a rod to an exhaust valve. The sensory subsystem is separable from the controller subsystem because the rod can be cut. Thus, this vessel meets his requirements of a negative feedback system.

It is not, however, obvious how the detachability requirement is supposed to eliminate my convective cooling counterexample. Sensory events and controlling events are, after all, merely certain designated events. They do not come labeled. Saying about a thermostat, "It *senses* the controlled property, the temperature, by means of a thermocouple or bimetallic strip . . . and it *controls* or corrects the temperature" (Faber 1986, 64), uses the terms of phenomenal experience to make a reductionistic analysis. Calling one kind of event "sensory" reads human perspective into the data and is acceptable only if its metaphorical status is kept in mind. It seems, therefore, as acceptable to designate as the sensory stage in the solar heating example the air's exiting the top of the air mass as to designate as the sensory stage the lifting of the saucepan lid or the moving of the piston, and it seems as acceptable to designate as the controlling stage the cool air's arriving at the bottom of the air mass as to designate as the controlling stage the lifting of the lid or the exhaust valve's opening. Further, the exiting hot air and the arriving cool air are connected by the air column between, similar to the piston and the valve's being connected by a rod. The hot air's leaving causally contributes to the cool air's moving in, which is similar to the piston's moving causally contributing to the exhaust valve's opening, so the sensory stage in the convective cooling example, as in the steam pressure example, causally contributes to the controlling stage. In the pressure vessel, the controlling stage was detachable from the sensory stage by cutting the connecting rod, while in the convective cooling example, the controlling stage is detachable from the sensory stage by blocking the upward movement of air. Although actually doing so would be unreasonably difficult with a large air mass, it is conceivable, which should be sufficient to meet the theoretical demand, but, in any case, is certainly manageable in convection cooling of that which is smaller, as stadiums and other buildings, or even trees and large boulders.

It is interesting that two of the diagrams of pressure vessels, one exhibiting negative feedback and the other not, differ only in the shape of the weighted relief valve. On one, the valve is a weight to the bottom of which is attached an extension flared outward, with the steam exiting around the circumference; on the other, the valve is a weight to the bottom of which is an extension in the shape of a cylinder, with the steam exiting the bottom as a hole is uncovered. It seems unlikely that negative feedback is captured by the slight difference in those valves. Indeed, if the first valve had a core drilled out of the center, it would then have both a skirted bottom and a cylindrical interior, and would bridge both categories. On the other hand, if negative feedback were defined merely as behavior that reduces variation and tends toward a certain position or value, it would amount to no more than a new name for the old concept of plasticity.

We all know roughly what a negative feedback system is and can cite clear examples. However, it has proved difficult to formulate the concept coherently and precisely. In spite of much effort, including Beckner's and Faber's formal analyses, questions remain. In the case of an electronic amplifier circuit, which seems to have been the origin of the concept, negative feedback has clear meaning, for a portion of the output current or voltage is returned as input with polarity reversed. Most people feel comfortable extending this to the concept of the common thermostatically controlled heating system, not because the polarity of the electric current is reversed but because an increase in heat, if great enough, causes its subsequent decrease. However, this is an extension of the original concept, for there is nothing fed back to the thermostat that corresponds to the reversed polarity current fed back in the amplifier. After all, coldness or lack of heat or even decreased heat is not fed back, but rather increased heat, which is the unaltered output of the system. Instead of concluding that this is a control system operating by positive feedback, negative feedback is saved by dropping reference to a reversed portion of the output being fed back to the input and describing the behavior in terms of output only, such as changing direction or reducing variation, as when Faber says, "Finally, in requiring that the feedback be negative, we stipulate that the system actively regulate or protect h_1" (1984, 84). Again, "suppose that some accidental rewiring of the temperature homeostat of a warm blooded animal turns that subsystem into a positive feedback loop, so that deviations from the normal temperature range are abetted rather than counteracted" (1986, 91) and "the device pushes the tiller to port when starboard is needed to restore the boat to its course. This would be an instance of positive feedback" (1986, 85).[8]

Falk

Like Adams and Faber, Arthur Falk looks to negative feedback to provide the key to understanding means-end relations and teleological behavior. Whereas they recognize only in passing the need for representation in feedback, Falk is emphatic about the necessity for representation to avoid the problem of reverse causation.

> Ends can play no role in their own genesis. Yet somehow they do. The solution is to say that at the beginning of the means-end sequence there

is, not the goal, but the thought of it. The thought of the goal can exist from the start, when the goal itself does not, and it can guide action to making that goal a reality. (Falk 1981, 199)

There is need, however, to find a way for lower organisms to be included, and that leads to the distinctive feature of his position, his theory of natural signs. "We must extend the notion of an unconscious thought of a goal state as well. For means-ends analyses to apply to unconscious organisms, they must possess *natural foresigns* of their as yet unrealized goal states" (1981, 199). A foresign is an internal state that represents the goal-state.

A foresign does not appear in nature by itself, but as paired with a feedback sign. The minimal system for natural signs is the negative feedback system (Falk 1981, 202), which he sees as undergirding teleology throughout nature, from the purposive behavior of humans to the pancreas' controlling blood sugar to a pole bean's growing around a fence. "All living organisms are negative feedback systems or systems of systems" (Falk 1981, 215).

A foresign operates by being compared to another sign, the sign of current output, the feedback signal. "The feedback about its own effects is also a sign" (Falk 1981, 202). So, there are two signs, the foresign and the feedback sign. Comparison is made between the two signs, not between the output and the goal-state.

To illustrate all this, Falk, like others, turns to the thermostatically controlled heating system. The foresign is the setting of the thermostat, but by that, unlike Adams, Falk does not mean the numbers on the dial. "It is not the numbers on its face which we read. That is the sign for us. What we have been referring to is the sign that is responded to by the thermostat-furnace system, not its translation into our symbol system" (Falk 1981, 201). In respect to a common variety of thermostat, the foresign is "the angular distance of a hand on a gear from a horizontal ledge that prevents the gear from further turning once the gear hand abuts it" (Falk 1981, 202). This distance, presumably, is significant because it affects the angle from the horizontal of the mercury switch.

The feedback sign is also a physical state. "Some physical state of the inner workings of the thermostat will be the sign for the system itself" (Falk 1981, 201). It is not further specified. Perhaps it is the curvature of the bimetal spring that changes with ambient temperature. In any case, the feedback temperature also affects the angle from the horizontal of the mercury switch, but in an opposite direction to the setting. As the setting is increased, the angle from the horizontal is increased; as the feedback temperature, or perhaps the curvature of the bimetal spring, is increased, that angle is reduced. The difference between the sign for the goal-state and the sign for the current temperature constitutes the degree of error. "The system reads its sign by responding to an error signal in a way that eliminates the error" (Falk 1981, 202).

Falk considers a foresign a protointention and a feedback sign a protobelief. He also sees the interaction of foresign and feedback sign as a kind of subtraction, with the foresign as the minuend and the feedback sign the subtrahend (1981, 202-203). In addition, he views the foresign as having an imperative mood, presumably because what it does is somewhat similar to ordering, and the feedback sign as having a kind of declarative mood, perhaps because what it does is somewhat similar to reporting (1995, 319).

The theory of natural signs in the context of negative feedback systems succeeds in preventing the problem of reverse causation. However, representation is also needed to account for the fact that there can be a goal that is not reached, requiring that the concept of goal not be tied to the reaching of it. It was noted earlier that this is a problem for any behavioristic analysis, such as a plasticity theory. It appears also to be a problem for a natural sign approach because of the need to have a purely causal account of interpreting signs.

Falk originally held that what a natural sign means or signifies is determined by the output of the system. "A natural sign does not signify any state of the system which the system does not interpret it to signify" (1981 202). This meant that the system could not misread a sign. Since it signified whatever the system response was, a natural sign also could not err in what it signified. "Signification and interpretation are interdefined in a way that precludes an independent criterion of the correctness of interpretations" (1981, 202). Because a foresign means whatever the system does, it cannot accommodate unsuccessful goal-directed behavior, such as fruitless search behavior, as an animal searching for water where there is no water. Furthermore, having both foresign and feedback signs errorless in signification and interpretation renders the theory unable to describe an excessive response (as an inordinately high fever in response to a minor infection) or an inappropriate response (as an allergic response to a harmless or even beneficial substance). Indeed, there is no way to describe a defective version of the primary example offered, the thermostatically controlled system. A bimetal spring in a thermostat might be faulty and assume the curvature it should have had for five degrees lower. Yet that could be determined only by comparison to a standard, which would involve going beyond the feedback system in question.

In Falk's most recent version, however, natural signs are made fallible. "The *feedback* is a sign too, a sign of conditions as they allegedly were when measured" (1995, 319). The two signs give "conditions as they allegedly just were with conditions as they allegedly should be" (1995, 319). Since they merely allege and do not guarantee, error is possible. In addition, the foresign is said to signify truly when it signifies a state that, for the organism, is good, that is, a state of fitness. If the organism migrates or if the habitat changes, a kind of error occurs. "When the zero state of an otherwise normal system is *not one of fittedness*—say it migrated—the system errs about its own good. Its foresign is false" (1995, 323). Error can also occur if the organism is abnormal. "Thus what's good for the healthy may be bad for the sick, as drinking water is for the dropsical. The foresign present in their thirst is in error" (1995, 324). Feedback sign error also occurs when there is "failure in the measuring or in the transport of feedback," as when a dunnock accepts a cuckoo's eggs (1995, 324).

The problem now is that since the connection between what a sign signifies and everything else, including earlier and later stages of the feedback system as well as the environment, is severed, there is no way to distinguish one sign from another or, indeed, if there is a sign at all. The zero error signal is supposed to help. "Foresigns signify states that, when they occur, cause a zero state of the error signal, and these states are generally good ends for the organism" (Falk 1995, 322). However, since nothing is error-free, the feedback signal or the subtraction process might also be in error. If a foresign is followed by a zero error signal when the organism is standing in water, that does not mean that the foresign was of water. It might have signified dry ground and been in error. Even if the foresign signified water, the feedback sign might have been in error, producing a zero error signal when the organism is standing on dry ground. Or error might have occurred in processing the feedback and foresigns, or in the corrective action taken, or in judging what is good.

For a sign to be capable of signifying incorrectly, there must be something it is supposed to signify. To provide for the normative element of signs, Falk introduces natural goods. The ultimate ones apparently are survival and reproduction (Falk 1995, 315) and subsidiary ones are those things that contribute to them (1995, 300-301), such as having food and water and rest, states of what he calls passive fittedness.

Even if such declared natural goods were granted, they would not provide sufficient variety and detail for the great range of things signified. Signs can be anything and can signify anything. Nothing about signs indicates how they are to be interpreted. Signs can signify conditions that are too fleeting ever to figure in natural selection as well as conditions that, having nothing to do with survival or reproduction, have no connection or relevance to it.

In several places, Falk speaks of his various concepts as having an analogical status: "To apply means-end analyses to the activities of all living things is to extend them by analogy from their base of proper application, human action" (1981, 199), and "The analogical use of means and ends, with its natural signs interpreted by natural processes, has been superimposed on the negative feedback model" (1981, 204). Regarding the reading of the signs, he says, "But part of the analogy I am recommending is to think of all living nature as literate enough to read and be guided by natural foresigns of its goal states" (1981, 200). The reference to subtraction is also meant analogically: "What the thermostat-furnace does may not look like anything you would want to call 'doing a subtraction problem'. . . . But the analogy to subtraction is sound" (1981, 202-203). Sometimes key terms occur in quotes, indicating nonstandard reading: "Part of the analogy is that living material is smart enough to 'read' the foresigns it contains" (1995, 317); also, "The two signs come together and get 'compared'"(1995, 319). The claim that the foresign is in the imperative mood apparently is meant analogically.

If negative feedback systems have features that are somewhat like signs but are not signs, perform actions that somewhat resemble the reading of signs, the giving of commands, and the reporting states of affairs, but are not those things, the value of referring to such features and actions to explain teleology is called into question. No one doubts that goals can be compared to all sorts of other things, including the angle from the horizontal of a component in a thermostat, but descriptions in terms of such comparisons fall short of explaining why the original is the way it is. It might be illuminating to compare swirling leaves to a dance, thunder to a threat, or flowing water to escape, but doing so does not explain why leaves swirl, thunder rumbles, or water flows. The analogue of the cause of an event is not a cause of that event, and in explanation we are usually interested in learning causes.

It is troubling that the bearers of natural signs are so difficult to identify. It is not at all obvious that the best candidate in the thermostat example is the angle from the horizontal of a projection on a gear.

Someone else might think it is the angle from the horizontal of the mercury switch. Adams thought it was the position of the dial. Moreover, the feedback sign was not identified at all. Thus, of the two natural signs in what is the most widely used example of a negative feedback system, whose parts and causal pathways are far clearer then those of organic feedback systems, only one is identified and that without unanimity. In one place he says the role of a sign is clear but that the sign token is AWOL (Falk 1990, 31). He notes that in Watt's steam engine governor it is not obvious what part of the system constitutes the foresign, then adds, "We can identify the features which function as signs, however vague their boundaries. Signs only become units the system can segregate and manipulate after minds evolve" (1990, 31). The fact that where natural signs have been identified, informed observers disagree in their identifications and the fact that some, perhaps many, natural signs cannot be identified at all should raise doubt about whether there are natural signs or, at least, whether the concept is clear enough to be useful. Saying that natural signs cannot be segregated and manipulated until minds evolve says that they cannot be segregated and manipulated until the natural signs become ordinary signs. That does nothing for the theory of natural signs.

It is difficult to bring functions under the mantle of feedback theory. Falk, like Adams and Faber, treats functions as parts or processes of feedback systems. "Only parts or processes of negative feedback systems of control have functions, and their functions are their contributions to realizing the systems' goals. Do our callouses and a rhinoceros's horns have functions? Yes. Do a frog's tongue-flickings for catching a fly? Yes" (Falk 1995, 332). To take care of things like tools, he adds, "Non-cybernetic artifacts have functions because we and they make a system." To block my charge that such a view requires saying that a branch has the function of supporting a bird because branch and bird constitute a system (Nissen 1993, 42), he adds that the item having the function must be made and not found (Falk 1995, 332-333). If this restriction means that the item must be made by man, then the functions of the parts and processes of organisms are ruled out. If the restriction means that the item must be made by the system of which it is a part, the functions of the components of a thermostatic heating system are ruled out. If the restriction is that the item must be made by some feedback system or other, not only are branches reinstated as having the function of supporting birds, but roofs and telephone lines are added. Requiring that something has a function only within a system, as we saw in the discussion of Adams and Faber, leads to bizarre and evanescent systems regarding which we have no names and apply no teleological predicates or describe in any way at all.

Falk describes himself as a semeiotic gradualist. He favors an approach that posits continuity from nonsigns to protosigns or natural signs to signs. There is a danger, however, that such an approach will dismiss genuine differences and discontinuities because of the belief that there are unknown linking increments in nature and that they explain the observed differences between signs and nonsigns. The increments might not be there. Some things in nature, like hurricanes and zephyrs, can be viewed as occupying locations on a continuum, but some things, even things so closely related as magnetic and electrostatic forces, cannot. Even if two things can be understood as positions on a dimension exhibiting minute increments or even continuity, it needs to be carefully examined what this implies for the character of the extremes.

In spite of the serious unresolved difficulties in his theory, no other analyst has been as clear and emphatic about the need for signs to explain teleology as has Falk. This marks a distinct advance in this long, even ancient, debate.

Summary

A feedback analysis certainly has its attractions, since much goal-directed behavior, perhaps most, utilizes feedback, and artifacts that mimic goal-directed behavior typically use it. Nevertheless, the many problems discussed above seem formidable. The first version, that of Rosenblueth, Wiener, and Bigelow, requires the goal-object to exist in order to send signals to the behaving device, rendering that account ill-suited for behavior lacking a goal-object, as in the case of unsuccessful search behavior. To remedy this, later versions require that the feedback system represent the goal-state. However, there is a remarkable looseness and uncertainty in identifying the component or property doing the representing and even in the case of a simple thermostat writers disagree on what that part is. It is equally difficult to determine what is represented. If that were determined by output, representation could not err and there would be no way for the device to fail. If not determined by output, it must appeal to norms. These norms must come either from within or without. If the norms come by a natural process from within, that needs to be explained. If the norms come from without, such as from natural goods, that needs some defense and, in any case, would undermine an explanation solely in terms of feedback.

It is not even clear which of the many causal consequences of a feedback system constitutes the output, even when these effects are restricted to equilibrium maintenance. A thermostatic heating system maintains many states besides that of a certain temperature in a building, such as the dimensions and rate of infrared radiation of many objects, and the insulin feedback system maintains much in addition to a certain glycogen level. Furthermore, in some behavior, as that motivated by extreme anger or fear, or by an organism with sensory deprivation, output may not be monitored by the system but still be goal-directed.

Functions fit awkwardly in feedback theory. Items that contribute to a system's output are said to have functions. However, if a function is whatever the item does that contributes to the output of the system, then malfunction is impossible, making even breakage, rust, and decay functions. If, to avoid that, one brings in context, usage, or design to determine functions, then the analysis has ranged beyond feedback. Further, hammers and doorknobs, pencils and paper clips are not parts of feedback systems. Extending the theory to include them by alleging enlarged systems composed of the item and the feedback system conjoined to or directing it, such as a person and a doorknob, produces unnatural systems, many of exceedingly short duration, concerning which we have no interest and do not speak teleologically or otherwise. Further, although parts such as the base or cover of a thermostat or the fur, teeth, and bones of organisms are parts of the feedback systems, they do not have their functions because of feedback and are not directed by a feedback system. In fact, being conjoined and directed by feedback systems does not suffice to give functions to natural features of the environment, such as oxygen and water, hollow logs and branches, and if adjusted to exclude them, give erroneous functions to artifacts, as barn roofs and telephone lines.

The intensionality of teleological concepts and extensionality of causation suggests that a purely causal analysis will not work. Finally, a feedback analysis needs to say something about the fact that knowledge of feedback is recent while teleological language is ancient and why teleological judgments are often, indeed, normally, made with no knowledge of internal structure.

Notes

1. It should not be assumed that the concept of negative feedback is either simple or clear. Some of the examples offered do not seem to exhibit feedback from the goal-object. Although a sonar-guided homing torpedo has feedback, a torpedo guided by the sounds of the ship's engines uses signals moving in one direction only, from the ship to the torpedo. There is, of course, feedback of a more limited kind within the torpedo, but the ship is not part of it. Wimsatt (1971) discusses some of the conceptual problems of feedback. The term "negative feedback" comes from electrical theory. Since electric signals may be positive or negative, it makes sense to speak of negative feedback in that context. Negative feedback is used most commonly in amplifier circuits, in which a portion of the output is returned to the input with the positive and negative parts of the waveform reversed. The authors are, therefore, wrong in saying (Rosenblueth et al. 1943, 19), concerning an electrical amplifier, "The feed-back is in these cases positive—the fraction of the output which reenters the object has the same sign as the original input signal." If, for example, a vacuum tube amplifier returned part of the plate current to the grid without polarity reversal, when the plate current is positive, the grid would become more positive, that is, less negative, than formerly. That would increase current flow from the cathode to the plate and, as the cycle repeated, the current would very quickly reach a maximum. If a component failed, the current would reduce to zero; if nothing failed, it would remain the maximum. In either case, the circuit would not amplify.

2. Engels writes (1982, 167), "Nun wäre es unsinnig zu sagen, dass der nicht mehr oder noch nicht erreichte Temperaturzustand es sei, der Signale mit dem Zweck seiner eigenen Realisation aussendet. Vielmehr ist es die Einstellung des Thermostats auf einen bestimmten Temperatursollzustand, der für die Zielrealisation verantwortlich ist."

3. However, Woodfield's position becomes blurred when he says, "All goal-directed systems operate by feedback" (1976, 197).

4. In a footnote, Adams says (1979, 495), "However, O may be a goal-state for S even if it is never reached." This seems, however, to play no part in the exposition of his analysis.

5. Found in Wimsatt (1971) and Beckner (1959).

6. However, sometimes Faber claims more. On one occasion he writes, regarding cybernetic theory, "The theory asserts that an organism or other system is oriented toward a goal state just if the parts are so arranged that

under certain circumstances it would act in a way that tends to make the goal state occur or more nearly occur" (1986, 61), and later, "When we ascribe a function . . . we call attention to the fact that the containing system tends to maintain some variable property . . . by means . . . negative feedback" (1986, 89).

7. Faber also says (1986, 97), "Can we produce a *general* analysis of teleological language, speaking with one voice about the application of teleological concepts to human beings, to their artifacts, and to natural objects? This is a question I have set outside the bounds of this book."

8. Faber does, however, also describe negative feedback in other ways. For systems in which the regulation is direct or instantaneous, as an a electric current regulator circuit, negative feedback is described mathematically in terms of the product of partial derivatives having a negative numerical value. For systems in which the regulation occurs gradually over time, as a thermostatic heating system, negative feedback is described in what apparently is a formal way of repeating the behavioral claim that the system decreases variability of a selected property (1984, 82).

Chapter 3

Natural Selection

The following sections on Canfield, Lehman, Ruse, Wimsatt, and Neander ground teleological language in biology on natural selection. Canfield does it in terms of probability of survival and reproduction, Lehman in terms of what he calls proper functioning, which seems to reduce to species survival, Ruse in terms of adaptation, Wimsatt, selection, and Neander, straightforward natural selection. Other authors also base their analyses on natural selection, at least in part, as Wright and Woodfield, whose works will be examined later.

Canfield

John Canfield notes that when scientists set out to learn the function of an organ, they often remove the organ and look for ways in which the organism is harmed. In the case of the thymus, it was not until the organ was removed from young animals that it was learned that its function was to produce lymphocytes in juveniles. Such a discovery procedure indicates, he reasons, that when we state what the function of an organ is, we state what the organ does that is useful to the organism. This reflection prompts his preliminary analysis of function statements.

> *FA* is a correct functional analysis of *I* *if and only if* *FA* has the linguistic form of a functional analysis, and *FA* states what *I* does in *S*, and what *I* does is useful to *S*. (Canfield 1964, 288)

"S" may refer to organisms, such as mice and vertebrates, or to subsystems of organisms, such as that regulating glucose, and it may refer to individuals as well as classes. "I" refers to organs, such as the heart and the liver, to components not usually considered organs, as leaves and roots, red blood cells and lymphocytes, and to processes, as the heartbeat and the secretion of bile. Like "S," "I" can refer to individuals as well as classes (Canfield 1964, 287).

In order to show that function language is acceptable in science because it can be translated into language that contains no objectional teleological terms, Canfield rephrases the above preliminary account as a preliminary translation schema.

> A function of *I* (in *S*) is to do *C means I* does *C* and that *C* is done is useful to *S*. (Canfield 1964, 290)

Explaining this, he adds, "For example '(In vertebrates) a function of the liver is to secrete bile' means 'The liver secretes bile, and that bile is secreted in vertebrates is useful to them.'"

There may, however, be some question about the term "useful." To sharpen the meaning, Canfield restricts the analysis to plants and to animals other than man and then defines "useful" in terms of preserving the life of the organism or the species (1964, 291). His revised and final translation formula reads:

> A function of *I* (in *S*) is to do *C means I* does *C*; and if, *ceteris paribus,* *C* were not done in an *S*, then the probability of that *S* surviving or having descendants would be smaller than the probability of an *S* in which *C* is done surviving or having descendants. (1964, 292)

The ceteris paribus clause is intended to exclude extraneous factors that might affect survival or having descendants, that is, factors other than C.

> The *ceteris paribus* clause assumes that the two specimens, e.g. the two mice, are: (i) alike in other relevant respects (i.e. respects other than *C*), and (ii) in other respects normal. The *ceteris paribus* clause also assumes that the two specimens are in environments which are: (iii) in relevant respects the same in both cases, and (iv) normal. (Canfield 1964, 291)

For example,

> A function of the heartbeat in mice is to circulate the blood; and if, ceteris paribus, the blood were not circulated in mice, then the probability of mice surviving or having descendants would be smaller than the probability of mice in which blood circulates surviving or having descendants.

Thus, Canfield's analysis correctly admits the function of the heartbeat being the circulation of the blood. With the simple verb "does," it avoids the problems of an analysis in terms of necessary conditions. We make function judgments about individuals as well as kinds, and this analysis provides for that. It excludes vestigial organs from having functions, since they would not increase the probability of individual or species survival. It avoids the difficulty feedback analyses have in accommodating the fact that we make accurate judgments of functions with no knowledge of internal structure. It has the virtue of being closely related to the procedures science uses in discovering the functions of organs and is a view with which many scientists are comfortable.

The ceteris paribus clause, as already noted, requires the compared specimens to be "in other respects normal." An otherwise normal mouse in which bile is not secreted will not have as high a probability of survival as an otherwise normal mouse in which bile is secreted. The normalcy requirement is included to rule out such cases as a laboratory mouse with a plastic tube sewn to the duodenum through which bile is injected.

It might seem, however, that the normalcy requirement rules out too much, for it appears to rule out a function of the liver being the production of bile in, for example, a blind mouse, since a blind mouse is not normal. However, the normalcy requirement is expressed in the form of a subjunctive conditional. It is not necessary that a specimen actually be normal to have functions, but only that if it were normal, certain consequences regarding survival and reproduction would occur. The normalcy requirement provides the standard conditions on which specifications of the analysis are stated, just as certain lighting conditions provide the standard conditions on which describing sulphur as yellow is based.

Nevertheless, Canfield's analysis has its problems. The most prominent is that it is an analysis of a limited range of teleological language. It leaves unaddressed the question of how function statements in biology relate to function statements about artifacts and to descriptions of goal-directed behavior. Wright observes, "He treats only the organs and parts of organisms studied by biology, to the exclusion of the consciously designed functions of artifacts. As a result of this emphasis, his analysis is, without modification, almost impossible to apply to conscious functions" (1973[a] 145). Even if one used the preliminary version of the analysis that describes functions in terms of what is useful, wider application fails, for, "if something is designed to do X, then doing X is its function even if doing X is generally useless, silly, or even harmful" (Wright 1973[a], 146).

Another reason why the analysis could not be extended to the functions of artifacts is that artifacts need not occur in systems. "In the conscious cases, there is an enormous problem in identifying the system *S*, *in* which *I* is functioning, and *to* which it must be useful" (Wright 1973[a], 145).

In spite of its title, "Teleological Explanation in Biology," Canfield's analysis does not cover all of biology. "Let us restrict the discussion to the biology of plants and of animals other than man" (Canfield 1964, 291). Humans are not included. The analysis is designed to handle the function of the heart in mice but not the function of the heart in humans! One reason for this surprising restriction may be that Canfield's preliminary translation schema analyzes functions in terms of what is useful to the subject, which his final version clarifies. Things useful to humans, of course, include much more than what contributes to survival. The analysis has an unfinished appearance because it does not extend to functions in humans.

Canfield considers as a possible objection an example of two bottle-capping machines (Canfield 1964, 292). They are alike except that one has an extra wheel that makes a screeching noise but does not contribute to the bottle-capping operation. It is possible, he says, that the machine with the extra wheel caps more bottles per hour than the other. The extra wheel has no function, yet the system of which it is a part is more successful than the other. He concludes, therefore, that the actual performance of the system cannot be used to determine whether the parts have a function. In response, he notes that the analysans is expressed as a subjunctive conditional. Although a machine having an extra wheel may accidentally cap more bottles, it is not the case that any such machine would cap more bottles. Expressing the schema in this manner is clearly a way to convey the idea that the part that has a function must contribute to success or reaching the goal.

It is, of course, odd to use as an example a machine, something that is both inanimate and a human artifact and so is excluded from the analysis. Presumably Canfield intends there to be a parallel extending to the context of plants and animals. Perhaps a strain of mice with an unusual pigmentation might illustrate his point. It might be that such mice accidentally survive better even though it is not in general the case that if mice had such pigmentation, they would survive better.

Lehman, however, argues that the subjunctive formulation does not succeed in doing what is needed.

The heart does produce a pulse in human beings and if *ceteris paribus*, a pulse were not produced in a human being, then the probability of that human being surviving or having descendants would be smaller than the probability of a human being in which a pulse is produced surviving or having descendants. (Lehman 1965[b], 327)

Canfield's analysis, therefore, would require us also to accept the statement, "A function of the human heart is to produce a pulse," which is false. Hence, Lehman concludes, the analysis must be wrong.

Canfield rejects Lehman's counterexample because if the heart caused the nourishment to reach the cells, survival would be unaffected by the presence or absence of a pulse. "That is, the heart, in achieving the transportation of nourishment to the cells, does not do so via the production of a pulse (though of course in achieving transportation a pulse is, as a matter of fact, produced)" (Canfield 1965, 330).

Canfield, thus, sees the pulse as outside the causal chain linking the heart to the transport of nourishment. This seems incorrect. The blood is moved by a hollow muscle contracting. Given this manner of propulsion, the flow will inevitably exhibit transient pressure variations occurring in step with the contractions. Canfield's statement, "We say $'I \rightarrow X. X \rightarrow Y'$, but not, e.g. $'I \rightarrow C. C \rightarrow X. X \rightarrow Y'''$ (where "I" refers to the heart, "C" to the pulse, "X" to the transportation of nourishment to the cells, and "Y" to some later unspecified result) is incorrect. Rather, what he says we do not say, "I $\rightarrow$ C. C $\rightarrow$ X," is correct. The pulse is an intermediate link in the causal chain, not a by-product or side effect.

Canfield further explains his position why having a pulse does not affect survival by noting that if an instrument were added that removed the spurts, survival would be unaffected. However, an animal with a pulse dampener would not meet the conditions of Canfield's ceteris paribus clause, for it would not be "in other respects normal," nor would the two animals being compared be "alike in other relevant respects." It would be physically impossible to have two organisms alike in all respects except that one had a pulse and the other did not. Removing the pulse would require making other alterations as well. Harry Frankfurt and Brian Poole express the point well: "An organism's activities arise out of its bodily structure, and its structure must be altered in order to change its activities. The notion of two organisms which have the same structure, but in which different activities take place, violates our ideas of causality" (1966, 71).[1] Once the two organisms differ in things besides the fact that one has a pulse and the other does not, the ceteris paribus clause is violated.

It might appear that the subjunctive form of the analysans would prevent the ceteris paribus clause from being violated as it did when talking about the function of the liver in a blind mouse. There the analysans applied to an abnormal S because it stated that if S were normal, the consequences would be such and such. However, here it is required that S be abnormal in respects other than the item being compared, for requiring that there be no pulse makes necessary other changes as well. Speaking subjunctively about a specimen being normal makes no sense if it is also required to be abnormal. Furthermore, there would still be a pulse between the heart and the pulse dampener; the pulse would be merely removed from some areas while remaining in others. One must, therefore, conclude with Lehman that a pulse does indeed meet the conditions of Canfield's analysis, even though the corresponding function statement is false.

Actually, any intermediate member of the action chain between the heart and the arrival of nourishment at the cells would serve as well as the pulse. Thus, C could be the moving of the blood through the ascending aorta, the left common carotid artery, the brachiocephalic artery, or the popliteal artery. Added to that, the components of the blood could be listed, such as the plasma, the platelets, the lymphocytes, and the eosinophils. Then, of course, one could generate another set by combining locations of blood flow with the components of blood, and talk, for example, about moving monocytes through the hepatic artery. The heart moves monocytes through the hepatic artery, and animals in which monocytes are moved through the hepatic artery have a higher probability of survival than do animals in which that is not done. Although this example meets Canfield's criteria, we would not say that a function of the heart is to move monocytes through the hepatic artery. These examples illustrate the problem that any analysis must face in handling intermediate members of the causal chain.

The analysis also fails to exclude entirely extraneous effects, such as heart sounds, and this for two reasons. As Frankfurt and Poole point out, heart sounds are, in fact, useful and have survival value.

> The usefulness of a biological item, in Canfield's sense, depends partly upon the environment of the organism in which the item occurs. Now the present environments of many vertebrates include physicians, and the practice of physicians involves making diagnoses which often rely on the character of their patients' heart sounds. The patients presumably have better chances of surviving and of reproducing if this diagnostic technique can be used on them than if, *ceteris paribus*, their health could

not be evaluated by listening to the sounds which their hearts make. (Frankfurt and Poole 1966, 72)

Thus, heart sounds meet Canfield's useful or survival test of functions.

Secondly, although, unlike the pulse, heart sounds are not part of an action chain that contributes to circulation, the heart physically cannot pump blood without also producing those noises. To remove the heart noises would require other changes, such as installing a sound dampener. This would, like the pulse dampener, violate the ceteris paribus clause, even with that clause expressed as a subjunctive conditional. Thus, Canfield's analysis incorrectly renders heart sounds as a function both because heart sounds contribute to survival and because they are a physically necessary side effect of the heartbeat.

The heart also cannot pump blood without emitting electrical signals. Therefore, the analysis incorrectly renders emitting electrical signals a function of the heart. Other counterexamples, such as removing glucose from the blood and flexing the pericardium, can readily be added. Because the heart physically cannot move blood through the arteries without doing these other things as well, if these side effects did not occur, the probability of survival would decrease, in many cases, with death instantaneous.

Canfield, in the ceteris paribus clause, also requires that the two specimens being compared be "alike in other relevant respects (i.e. respects other than C)" (1966, 291). Thus, in comparing two mice in respect to bile secretion, it is not required that they be the same color, but it is required that they both have intestines. However, in determining this we must already know that the color of the skin and hair does not affect the role of bile in the life of the organism, that bile secretions flow into the intestines, and that, therefore, bile secretion without an intestine would be without survival value. Thus, to know that color is not a relevant similarity and that having intestines is a relevant similarity, one must already know how bile secretions contribute to the life of the organism. One must know the results of Canfield's test in order for the test to be administered, and if the results are already known, the test is not needed. The appeal to relevant features of the two specimens, thus, involves a circularity that renders the test either impossible to perform or useless to perform.

The ceteris paribus clause also requires that the environments be "in relevant respects the same in both cases" (Canfield 1964, 291). Suppose the function statement being considered were "A function of fur on rabbits is to reduce heat loss." By selecting a warm environment for clipped rabbits and an extremely cold environment for furred rabbits, it is possible that the experiment result in furred rabbits having a lower probability of survival. Presumably the requirement that environments be relevantly similar was intended to forestall such misleading test results.

Nevertheless, the problem that arose in respect to relevantly similar specimens arises again in respect to relevantly similar environments. To know that, in the above example, temperature is a relevant property and level terrain is not, one must already know what fur does for rabbits. If one does not know that, the test is impossible to perform; if one does know that, it is useless to perform.

The normalcy requirement cannot perform its intended role, for it says that the specimens compared must be normal only in those aspects other than the ones being compared. It is necessary that they need not be normal in the aspect being compared, for the point of comparing them is to see whether lacking the item affects survival. This means, however, that the gate to abnormal, even bizarre, conditions is opened wide. Any abnormality or freakish condition that happens to contribute to survival and reproduction is acceptable, such as a benign tumor that protects the kidneys from shock, a third kidney, an extra digit, and so on.

The ceteris paribus clause also requires that the environment be normal. We have some idea of what a normal mouse or rabbit is. What constitutes a normal environment seems intolerably vague. Nevertheless, it rules out too much. A laboratory, presumably, is not a normal environment for any animal; yet it is there that functions are commonly and successfully investigated. Interpreting the ceteris paribus clause as requiring only what is relevantly normal, thereby reinstating laboratories, would assume knowing which conditions affect survival in order to be tested for affecting survival.

There is a certain course-grained feature to function talk in biology that Canfield's analysis does not capture. It is a function of the eye to see, not to see either a certain kind of object, such as trees or water, or a specific object, such as this tree or that stream of water. Although it is true that the eyes of a mouse enable it to see calico cats and that if, ceteris paribus, it did not see calico cats, the probability of its surviving or having descendants would be smaller than the probability of a mouse that sees calico cats surviving or having descendants, the corresponding function statement is not true.

Richard Sorabji argues that functions cannot be defined entirely in terms of survival because there are other goods that can ground functions.

Suppose that an organ were discovered which came into operation only when some lethal type of damage had occurred, e.g., a major coronary thrombosis. And suppose that the effect of this organ were to shut off sensations of pain as soon as such lethal damage had occurred. This effect would not increase the chances of survival either for individual or for species. But it would confer a good of another sort. For the shutting off of unnecessary pain is a good. And it would be perfectly correct to say that the function of the organ was to shut off pain when lethal damage had occurred. (Sorabji 1964, 293-294)

He calls such a function, following a distinction made by Plato and Aristotle, a "luxury function." Although there is no such specific organ, there is evidence that the brain does release powerful hormones that block pain in cases of severe trauma. Such hormones are usually interpreted, perhaps somewhat tentatively, as having the function of blocking pain. Not only does such blocked pain not contribute to survival but it may, on occasion, even contribute to death. Descriptions of accidents sometimes include an account of a severely injured person neglecting medical aid until it is too late simply because the person felt no pain. Even when the injured lives, the blocked pain sometimes allows continued vigorous activity, exacerbating the injury.

As mentioned earlier, Faber imagines a mutant earthworm in which the rudimentary heart is replaced by cilia. The cilia maintain circulation of internal fluids, but do so less efficiently than did the rudimentary heart. The mutants have a lower survival rates; so the cilia are maladaptive.

Nevertheless, it is easy to see that promoting the circulation of internal fluids is a function of the cilia—not because this circulating action produces an increased survival rate (i.e., not because of selection for this trait), but simply because it leads to the continued existence of the individual organisms which possess the trait; without it the mutant worms would die. (Faber 1984, 95)

Surely Faber is correct in his conjecture that the cilia would be given functional status, contrary not only to Canfield's analysis, but to any analysis that appeals to improved survival and reproductive success to ground functions.

The analysis says that I does C and that doing so improves the rate of survival and having descendants. Yet we also know that vast numbers of species have become extinct in the past and that the extinction process continues today. In the process of extinction, doing C no longer improves the rate of survival or having descendants, yet we continue to assign the old functions. Where ground cover or a specialized food supply vanishes due to the expansion of agriculture, camouflage coloration or specialized organs of ingestion and digestion do not improve survival rate. Canfield correctly does not exclude extinction conditions, for that would trivialize the analysis. It would be in error to claim that extinction conditions should not count on the grounds that during those periods the environment is abnormal because, since extinction is constantly going on, that would render all periods abnormal. One could make the concept of normal environment normal relative to a species, saying it is normal when the species flourishes and abnormal when the species declines, but that would remove any explanatory usefulness of the concept of a normal environment.

Lehman

Hugh Lehman, like Canfield, addresses only the problem of function statements in biology and makes no attempt to relate such statements to teleological language generally. He offers his analysis in two slightly different forms. The form appropriate for structures, such as organs, reads:

> A function of X is Y = df. There is at least one organism such that: (a) some instance of X is a part of the organism and is not a sufficient condition or cause of malfunctioning of the organism and, if anything is an instance of X then its activity causes some instance of Y (providing, of course, that the conditions necessary under normal circumstances for instances of X to cause instances of Y are satisfied) (b) it is a necessary condition for the proper functioning of the organism that it exhibit some instance of Y. (Lehman 1965[a], 12)

He offers as an example, "A function of the heart is to circulate the blood." Following the pattern above, his analysis would be,

> There is at least one organism such that: (a) some instance of a heart is part of the organism and is not a sufficient condition or cause of malfunctioning of the organism and, if anything is an instance of a heart then its activity causes some instance of circulation of the blood (providing, of course, that the conditions necessary under normal circumstances for instances of a heart to cause instances of circulation of the blood are satisfied); (b) it is a necessary condition for the proper functioning of the organism that it exhibit some instance of circulation of the blood.

A slightly modified version is provided for function statements about processes, such as hunger or breathing (Lehman 1965[a], 13).

The analysis has been carefully thought out and has a number of interesting features. The unusual stipulation that X does not cause malfunctioning is included because of consideration of the ability of an organism to repair itself. A nonfatal blockage of the coronary blood vessels causes the growth of new vessels and sometimes leads to recovery. Nevertheless, Lehman notes, we would not say that the blockage had a function even though such blockage is a part of the organism and causes collateral vessel growth that is necessary for the proper functioning of the organism (1965[a], 14-15).

Instead of beginning with "For all organisms such that," Lehman begins with "There is at least one organism such that" because of the problem of vestigial organs. "Consider the case of vestigial organs such as the wings of such flightless birds as chickens and ostriches. While we would say that a function of wings is to enable birds to fly, it is not true of all properly functioning organisms with wings that the wings enable the organism to fly" (1965[a], 12-13). How to handle vestigial functions, however, seems unclear and many feel that the wings of flightless birds such as ostriches and penguins have lost their function.

Failure, such as a heart's beating without circulating the blood as in the case of a severe wound, is provided for by the normalcy condition. "The class of hearts can have this function even though particular heartbeats are not circulating the blood because some condition which is necessary under normal circumstances for the heart to circulate the blood is not satisfied" (Lehman 1965[a], 15). Although he did not mention it, goal failure also seems provided for by the phrase, "There is at least one organism such that."

Lehman separates the relevant effects from the irrelevant, for example, circulation from heart sounds, by restricting effects to those necessary for proper functioning of the organism. "The solution of this problem involves the idea of 'proper functioning' in our definition. By use of this idea we have tried to eliminate the problem of 'intuitively unpurposive' activity" (1965[a], 15).

There are, however, problems with Lehman's analysis. The condition that X, to have a function, must not cause malfunctioning is too strong. Hearts pumping blood cause strokes and gastric secretions cause ulcers; yet both hearts and gastric secretions have functions. Woodfield notes, "Nor is there good reason to stipulate, as Lehman does, that an item having a function must never do anything that conduces to the malfunctioning of the organism. A good kidney may cause damage to a blocked bladder simply by functioning properly" (Woodfield 1976, 114).[2]

The fact that Lehman regards the expression "There is at least one organism such that" as providing for a function of X to be Y even though X's activity does not always bring about Y means that X may have a function Y even though only one organism meets the conditions. After all, this is how he made provision for vestigial organs to have functions. This interpretation is too broad, for it requires that a rare abnormality that happens to improve proper functioning be given the status of having a function. A benign tumor covering the kidneys might result in a measure of trauma protection, making it acceptable, on Lehman's account, to say,

> There is at least one organism such that: (a) some instance of a benign tumor covering kidneys is part of the organism and is not a sufficient condition or cause of malfunctioning of the organism and if anything is an instance of a benign tumor covering kidneys then it causes some trauma protection . . . ; (b) it is a necessary condition for the proper functioning of the organism that there be some instance of kidney trauma protection.

Although most humans manage to function properly without such kidney protection, condition (b) would be satisfied if the organism would function better with such protection than without, such as people subject to severe and frequent shock, as players in violent sports. A single person having a sixth digit would be another example. It would improve grasping ability and, therefore, improve proper functioning. In general, any kind of abnormal condition or property or process that improves proper functioning in at least one organism would be a counterexample.

The analysis is also deficient in that it does not provide for the fact that not all effects necessary for proper functioning are functions, as in the case of the old and familiar example of the nose supporting eyeglasses. The example fits the terms of the analysis, for it is certainly true that there is at least one instance of a nose in which the nose is a part of the organism and does not cause malfunctioning of the organism but does provide support for eyeglasses, and, further, that supporting eyeglasses is necessary for that organism to function properly. For many, wearing eyeglasses is essential for good vision and good vision is certainly necessary for proper functioning. Nevertheless, we would not say, "A function of the nose is to support eyeglasses."

Lehman is aware that something can have a function without some of its consequences being necessary for proper functioning of the organism.

"The function statement does not imply that any particular items are necessary for the proper functioning of the organism in which they occur. For example, circulation of a particular volume of blood is not necessary for the proper functioning of some organism" (1965[a], 12). However, he then adds that the effect of the item having the function, that is, the function, itself, is necessary for proper functioning. "What is necessary for proper functioning is that the general condition 'circulation of the blood' be instantiated, that is to say, that blood be circulated in the organism." This is expressed in his condition (b). However, making the function, that is, Y, necessary for the proper functioning of the organism, is too strong. The palatine tonsils have the function of filtering out and destroying invading bacteria in the throat, but that consequence of the tonsils' activity is not necessary for proper functioning. The spleen has the function of destroying old red blood cells, but that consequence of the spleen's activity is not necessary for proper functioning. Even if Y causally contributes to a later state, Z, which is necessary for proper functioning, such as protection from invading bacteria, there may be alternative ways of producing Z. In that case, Y can be a function (of X) without being necessary for Z.

Another problem concerns the concept of proper functioning. Lehman explains, "To say that an organism is functioning properly is to say that its activity satisfies a certain standard or norm" (1965[a], 16). Such norms, he adds, have nothing to do with what we approve or whether the organism has intentional states. The concept of proper functioning does not imply that the organism is a means to some person's end or that there are purposes in nature. "So far as we are concerned, the standard exists if among persons with sufficient knowledge of the organism in question there is general agreement as to criteria of proper functioning" (1965[a], 18).

Saying that there will be agreement among knowledgeable people concerning the standards does not carry the discussion far. If the knowledgeable people use a criterion in reaching agreement, that criterion becomes the determining factor, not the people who use it. If no criterion is used, then proper functioning could be anything the people agree on, including the bizarre and unethical. If, to prevent that, the term "knowledgeable" is so understood that observers are not considered knowledgeable if they agree on such standards, then the standard vanishes in circularity.

As something of an afterthought, Lehman asks whether there is any reason for the agreement regarding proper functioning and responds by saying that it seems that "activities which constitute proper functioning for an organism are those which give the best likelihood, in certain environmental conditions, of its survival or of the continuance in existence of the species of which it is a member" (1965[a], 18). Since this supplies a criterion for proper functioning, it renders earlier remarks about what sufficiently knowledgeable people agree on irrelevant. Talk about proper functioning becomes unnecessary, for that is now understood as that which tends toward species survival, and reference to species survival would have been sufficient.

Lehman thus espouses at different times three different positions regarding the end or goal-state used to identify effects that are functions from effects that are not proper functioning, leaving the concept undefined; proper functioning as determined by whatever knowledgeable observers agree on; and proper functioning as that which tends toward species survival. The first one, which is the one appearing in the formal analysis, makes that analysis unworkable, since there is no way to determine when proper functioning has been observed. The second one, proper functioning understood as what knowledgeable observers agree on, either allows unacceptable, even repugnant, states to be proper functioning or is circular. The third one, whatever contributes to species survival, is the one conventionally given, but it is unclear what status Lehman gives it, since it does not appear in the formal analysis. Perhaps this is the one Lehman really has in mind and would like the reader to take his analysis as being grounded on natural selection.

Not addressed in Lehman's analysis is the relation of function statements in biology to function statements in other areas, including artifacts, and the relation between functions, goals, and goal-directed behavior, in short, all the rest of teleological language.

Wimsatt

William Wimsatt offers an unusually detailed and comprehensive examination of function statements. It is not an analysis of all areas of the teleology, but is restricted to functions, primarily, biological functions. He defends, as have the others studied, the position that reference to functions can safely be made in the life sciences without bringing along objectionable implications. Function statements make "no appeals to 'backwards causation,' 'vitalism,' 'entelechies,' or anti-reductionist sentiments of any sort" (1972, 5). He summarizes:

The end result, I hope, was to show that functional analysis and explanation are scientifically objective and respectable modes of analysis; that they can have a *bona fide* theoretical status as a means of applying theories of a certain logical form (involving the presence of differential selection processes); and that teleology, properly so called, does have a respectable role in the scientific characterization of non-cognitive systems. (1972, 80)

He specifically rejects the analyses of Braithwaite, Nagel, Hempel, Lehman, Sommerhoff, Ashby, Fodor, Putnam, Beckner, and Rosenblueth, Wiener, and Bigelow. In the course of doing this, he denies the claim that functions are merely causal consequences (1972, 5), as well as the opposing claim that causal consequences are irrelevant (1972, 10-11). Wimsatt singles out for special rejection the necessary condition analysis: "It is interesting to note that a concentrated study of examples of vital functions is probably responsible for the frequently held but erroneous view that functional things are indispensable or necessary for the survival of an organism, system, or species" (1972, 9). He also repudiates the negative feedback and the self-regulation analyses (1972, 8-9) and rejects the entire attempt to translate function statements into nonfunction statements, saying, "Most of the extant analyses—which take as their aim the translation of teleological statements into 'equivalent' . . . non-teleological ones—are doomed before they start" (1972, 66).

Wimsatt's procedure in seeking the correct understanding of functions is to determine what is needed to identify a function, supplying examples that would give only ambiguous support for a function claim unless the feature in question is taken into account. Instead of looking for a translation formula, his aim is to find, in effect, a decision procedure for identifying functions. We will understand what functions are when we understand how functions are identified. He never, however, makes it clear how the final product will differ from a translation formula, other than in grammatical or formal features.[3]

Identifying a function requires first indicating that which has the function. If it is an object, either material or abstract, it is called an item, abbreviated "i." Examples are the liver, a fountain pen, and a political party (Wimsatt 1972, 19). If that which has the function is something dynamic, a process, it is called a behavior, abbreviated "B." Examples include the circulation of the blood, the motions of machines, and human action (1972, 26-27). Wimsatt regards the category of behavior as primary and the category of item as secondary. Indeed, he regards mention of the item as redundant because an item will produce consequences only by action (1972, 26-27). Wimsatt regards the category of behavior as primary and the category of item as secondary. Indeed, he regards mention of the item as redundant because an item will produce consequences only by interacting with other parts of the system, that is, by its behavior. The lesser status accorded to the item is presumably indicated by "i" being in lowercase and its redundant and unessential status by being enclosed in parenthesis, thus "(i)" (1972, 29-32).

A given item or behavior may have more than one function. To identify each function, reference must be made to the system, indicated by "S." Thus, a function of peripheral capillaries in higher mammals is to control heat loss within the thermoregulatory system but is to allow the exchange of nutritive and waste materials between the blood and the cells in the metabolic system. Wimsatt regards it as important to carry the analysis down to the point of functional uniqueness such that one item does not have more than one function, and the reference to system helps do that (1972, 19-20). "Nonetheless, *assuming* functional uniqueness leads to a discovery of the conceptual variables upon which attributions of functionality rest much in the same way as the assumptions of causal determinism can lead to the discovery of the physical variables" (1972, 18). A single item or behavior might have different functions in different environments. Thus, the swim bladder of lungfish has the function of aiding in the perception of rate of climb or of diving when the environment is water, but it has the function of allowing oxygen intake when the environment is air. Wimsatt, therefore, adds variable "E" for "environment" as one of the determiners of functions (1972, 20).

The claim that circulating the blood is a function of the heart is acceptable, but the claim that making heart sounds is a function of the heart is not. However, the difference is not adequately explained, says Wimsatt, by merely noting that one aids survival and the other does not. Circulation aids survival only because the organism has a certain structure and because certain laws hold. If the organism had an acoustic homeostasis control mechanism that was mediated by sounds the organs made and if the theory were accepted that blood is an efficient conductor of sounds, then the claim that a function of the heart is to produce heart sounds would be quite acceptable. This shows, says Wimsatt, that functions are identified relative to background theories, and, therefore, the variable "T" must be added (1972, 28-29).

Wimsatt considers a servomechanism that, when radiation is detected, moves toward the source of the radiation, connects to it, and recharges its power supply. Charging time is limited to a certain number of seconds and charging rate is proportional to the strength of the charging source. It is possible, however, for overcharging to occur, in which case the power supply would be destroyed. The power sources are of varying strength, with some strong enough to destroy the power supply. The machine wanders about, moving from one power source to another. Sooner or later it encounters a lethal source, and its power supply, and, therefore, the machine, is destroyed (Wimsatt 1972, 20-21).

The behavior of the servomechanism is consistent either with the function of the machine being to keep itself charged or to destroy itself. The behavior is also consistent with the function of its charging limiter being either to prevent overcharging or to avoid wasting time on power sources that are too weak to destroy the machine. Thus, in both cases, knowing the behavior is not sufficient to enable one to discover the function. Even knowing, in addition to the behavior, the item, the system, the environment, and the background theories, one still is not able to determine the function.

> This example has a very important point: the fact that we could not infer the function of the limiter merely from a knowledge of the environment, causal operation, structure, goal-objects (the power sources), and behaviour of the machine is sufficient to show that attempted analyses of purpose or function in terms of goal-directive behaviour or of the structure of a self-regulating system (as advanced by Rosenblueth, Wiener, and Bigelow, Sommerhoff, Ashby, Beckner, Nagel, and Braithwaite) cannot succeed. (Wimsatt 1972, 22)

What is lacking, according to Wimsatt, is knowledge of the purpose of the designer. "The plans, internal structure, and behavior of this device are consistent with two mutually contradictory purposes on the part of its designer" (1972, 21). Wimsatt generalizes from this example to the claim that the determination of function requires knowledge of the purpose, abbreviated "P." That purpose is a major part of this analysis is clear when he says,

> It must be appropriate to speak of a purpose if function statements in the teleological sense are to be legitimate. Where purposes are appropriate, so are teleological explanations. (Wimsatt 1972, 62)

The kinds of knowledge needed for determining functions are now in place. The symbolic form of what he regards as a kind of decision procedure for functions is "F[B(i),S,E,P,T] = C" (1972, 32).[4] It is read "A function of F of behavior B of item (i) in system S in environment E relative to purpose P according to theory T is to bring about consequence C." It is used as follows:

> First choose an item for investigation. Then pick out the things which it does as a part of a given system. This determines the values of *B* and *S*. Then, using the laws of causal theory *T*, determine all of the consequences this behaviour of *i* has for *S* in a given environment, *E*. Finally, from these consequences pick those which result in or contribute to the attainment of purpose *P* by system *S*. These consequences are the functional consequences. (1972, 29-30)

The analysis is unusual in its complexity, for Wimsatt has no fewer than five factors determining functions. It is especially unusual in that he includes reference to purpose while denying the implication of mind or design. Indeed, the role of purpose in his analysis is greater than he acknowledges, for one must already know (i) or B in order to ask what the function of something is, and if one knows P, one can identify the function without knowing S, E, or T. This is most clearly the case with artifacts. If one asks about the function of a scythe, (i) is already identified, and if one learns that its purpose is to cut grain, one does not need to know what system or environment the scythe is used in or any theory of cutting.

Usually having an analysis utilizing purpose automatically takes care of the problem of malfunctions. However, in this version, the function is found among the consequences of (i) or B. This means that if the consequence that fits purpose P does not occur, (i) or B has no function. To meet this problem, Wimsatt suggests that functions apply only to types. "Such entities are said to have a certain function in spite of their failure to perform functionally because that *type* of entity has that function" (1972, 47). Elsewhere he suggests a probabilistic solution: "An entity could be regarded as functional if its presence or operation produced an increase in the probability of purpose-attainment" (1972, 55). He also says that the consequent will occur (1972, 5), that the consequent must occur only in the most inclusive system (1972, 49), and that most and probably all functions are highly successful. "Certainly most and probably all functional relationships involve relatively high frequencies of performances of the functional consequence" (1972, 50). Even such a wide range of differing answers does not suffice, for artifacts sometimes have functions that always fail, including some that cannot succeed. Further, biological function statements are often made about individuals, including individuals for whom success is not only unlikely, but impossible.

Most analysts have avoided placing purpose in the analysis precisely because it is generally felt that sharp separation of purpose from mind is impossible. C. J. Ducasse expresses something close to the common view when he says, "Only the acts of entities capable of belief and desire, are capable of being purposive" (1925, 154). Wimsatt endeavors to effect the needed separation in two stages, first by identifying mind with consciousness and then by separating purpose from consciousness by analyzing purpose in terms of selection.

It is clear that he implicitly identifies mind with consciousness, for from the observation that purpose is separable from consciousness, he implies that purpose is separable from mind. Otherwise his analysis would leave function statements with the objectionable implications he says they do not have.

> That is, I claim that all of the important logical features of purposiveness in the human case are also to be found in cases where consciousness is not presupposed or implied, and further, that consciousness, *per se*, is a detachable implication of statements involving purpose or teleology in their explanatory roles. (Wimsatt 1972, 12)

There is a preliminary difficulty about his statement that consciousness is a detachable implication of statements involving purpose. In analyzing the meanings of statements, what a statement implies is surely impossible to deny. "The cat is a mammal" implies that the cat is an animal and one cannot assert the first and deny the second. Implications can be ignored for practical reasons, of course, but logic, not to mention communication, would collapse, if one could assent to a proposition while withholding assent to its implications. If a statement ascribing purpose implies reference to consciousness, then the reference to consciousness is not detachable and cannot be denied.

However, a statement ascribing purpose does not imply reference to consciousness. It is not that the implication is detachable, but that there is no implication to be detached. The point does not seem controversial and has been made by many over the years. Thus P. H. Nowell-Smith says, "Neither intention nor purpose needs to be in mind before the action is done" (1960, 98). J. L. Cowan makes the same point, "The agent himself may not be aware of his purpose" (1968, 320), as do Carl Hempel, "In many cases of so-called purposive action, there is no conscious deliberation, no rational calculation that leads the agent to his decision" (1962, 29) and Jonathan Cohen, "Moreover, although many purposive actions are deliberate I think there are also many which are not. To describe a purposive action as deliberate entails asserting that the relevant dispositions were actualized in certain occurrent thoughts as well as in the action itself" (1951, 287).

Some actions are performed too rapidly for conscious purpose to occur. Borrowing an example from Nowell-Smith, "A man who grabs at a child to stop him falling over a cliff does not form the intention of grabbing; he just grabs; but his action is both intentional and purposive, since he could properly say 'I grabbed the child to stop him falling over the cliff'" (1960, 98)

Other actions that are frequently unconsciously purposive are habitual actions, such as tying one's laces or washing one's hands and are often performed with attention directed elsewhere. Cohen again: "But when I utter the phrase 'Twopenny one, please' on a Dundee tram going into town in order to buy my ticket I often do so automatically and with my thoughts on other things. And I think that a great deal of human talk and other conduct is unconsciously purposive in this way" (1951, 287).

Yet a third class of actions generally considered purposive but without those purposes being conscious is made up of actions that are components of larger, more complex actions, as the action of walking in the larger action of answering the door or the action of placing one's foot on the brake in the larger action of stopping at a stop sign. Purpose does not require consciousness; having a purpose does not require being conscious of that purpose.

Wimsatt, in another passage that might be intended to show that purpose does not require mind, says,

> Freud's attack upon explanations of action in terms of conscious purposes is now well known. One may doubt his or other psychological theories, but they cannot be thrown out on *a priori* grounds and acceptance of them leads to uncertainties concerning the purposes for which various actions are performed. (1972, 65)

If, as Freud says, the real purpose of an action may be different from the conscious purpose, then all we can conclude is that there are unconscious purposes. Establishing that purpose does not presuppose consciousness does not establish that purpose does not presuppose mind. To believe that it does, one must incorrectly identify mind with consciousness. As mentioned earlier, the general reluctance to analyze teleology and functions in terms of purpose is presumably due to the suspicion that purpose does presuppose mind. One must conclude, therefore, that since identifying mind with consciousness is incorrect, arguing that purpose does not presuppose consciousness does not at all establish that it does not presuppose mind.

If it could be shown that something other than mind, something that does not presuppose mind could account for purpose, the commonsense link between purpose and mind would perhaps be weakened. Wimsatt believes that selection can do just that: "The operation of selection processes is not only *not* special to biology, but appears to be at the core of teleology and purposeful activity wherever they occur" (1972, 13). Again, "Further, where teleological explanations are appropriate there is always (at least as a matter of empirical fact, and perhaps more) a background of the past operation of selection mechanisms to produce functional or purpose-directed organizations in the functional system" (1972, 62).

Selection occupies a far greater place in his theory than one might think in view of the fact that it does not appear in his formula. It is selection that ties purpose in human action and the function of artifacts to biological functions—in the former conscious selection, in the latter, natural selection. Wimsatt sees the variables P and T connected in that background theories are needed to identify purposes, and sees evolutionary theory, with its emphasis on natural selection, as the relevant background theory for identifying purposes in biological functions. "Thus, the purpose associated with biological adaptive function is to be derived from the structure of evolutionary theory—in particular from the mode of operation of natural selection" (1972, 63).

Since Wimsatt explains function in terms of purpose and purpose in terms of selection, it is essential that he explain selection in such a way that it does not require mind. Although the concept of selection is, thus, the cornerstone of his analysis, he says very little about it, and what he says seems incorrect or unhelpful. "In each case there are two correlative processes involved in selection, aptly named 'blind variation' and 'selective retention' by Campbell. . . . The concept of progress through trial and error is virtually synonymous with 'blind variation and selective retention'"(1972, 14).

Some selection behavior doubtlessly is connected with blind variation and selective retention, such as maze learning. However, even here, strictly speaking, the blind variation occurs before the selection and is not part of selection, itself. That leaves only selective retention, and to explain selection in terms of selective retention is circular. Whatever questions there are about selection are also questions about selective retention. Michael Simon observes, "To describe what the mouse is doing in terms of trial and error is to employ a teleological mode of discourse" (1971, 191).

Wimsatt's explanation of selection can be summarized as saying that selection is composed of blind variation and selective retention. Exceptions are all around us. From a shelf of books, one might, with a complete absence of false moves that could be considered blind variation or trial and error, select a particular volume. It is more common to see predators select prey with no trial and error than with it.

Wimsatt responds to such a challenge by adding that the trial and error might be mental. "In many cases of human problem-solving, the trials are *mental* trials or *Gedankenexperimente*, where the only physical trial that occurs is the chosen alternative" (1972, 14). Although one selects the correct book with no overt trial and error, there occurred mental trial and error. Although the clerk makes the correct change with no discernible missteps, there were mental missteps.

The occurrence of mental trial and error is claimed without supporting argument. No doubt they sometimes occur, but whether they occur frequently enough in ordinary affairs to do the job Wimsatt needs done is doubtful. Apart from that, it is singularly inappropriate in claiming that function statements do not presuppose mind to appeal to mental events in the analysis. Mental trial and error certainly presuppose mind, and appeal to them undermines the argument to establish purpose as mind-independent.

Another way that Wimsatt offers to support the claim that all selection involves trial and error is to allow the trial and error to have occurred at an earlier stage. "Finally, while there are many purposive actions which may not involve (on *that* occasion) either mental or physical trial and error, these are generally held to be explicable in terms of already knowing what to do because one has done it before; habit, conditioning, or instinct" (1972, 14). Selecting a certain book from the shelf may not involve trial-and-error behavior at the time of selection, but it is linked to many and diverse trial-and-error episodes ranging back into early childhood in learning to walk, to read, and to grasp objects.

Blind variation and selective retention and progress through trial and error were, however, offered as a way of explaining what selection is. They were not offered as mere preliminary or necessary conditions. A current act of selection is no doubt related to past learning, and that learning likely involved trial and error, but the components or structure of an act of selection can hardly be things that occurred at a different time. The constituents of a thing must be contemporaneous with that thing. An account of what something is is different from an account of how it got that way.

Selection has a specificity or uniqueness that distant prior trial-and-error episodes lack. The librarian selects, not just any book, but this book. The clerk selects, not just any coins, but these coins. Whatever relevant prior trial-and-error episodes there might have been would relate only to learning general skills, such as learning to walk, read, and handle objects, and would be relevant only to selection in general.

Wimsatt acknowledges that an account of selection understood as blind variation and selective retention will not serve as an analysis of teleology. A being could have genuine purpose and, hence, its actions could be explained teleologically, without having a history of blind variation and selective retention.

> In other words, while it might even be a law of nature that all teleological systems and processes result from some species of 'blind variation and selective retention', we at least conceive it to be *logically* possible that teleological explanations are appropriate to the behaviour of, and purposes are possessed by beings or systems which have never felt the moulding force of, these two processes. (1972, 15)

While he allows the logical possibility of teleological behavior that does not result from selection or blind variation and selective retention, he contends that selection is sufficient for such behavior. On the contrary, it is apparent that one can select or choose things for all sorts of reasons, including reasons that do not confer functions. Selecting the best dog in a dog show or the reddest apple in the basket does not give the dog or the apple functions. Although the act of selecting is purposive, its products need not be.

Adams argues that natural selection is not necessary for biological functions because a function may change without a new selection occurring. He offers as an example pearlfishes, whose behavior of inhabiting the sea cucumber by day and emerging at night to feed was originally selected for as a defense against predators. Later they lost eye pigmentation and now the function of the behavior or inhabiting the sea cucumber is to protect against light. He adds that vestigial organs indicate that selection is not sufficient for functions, for they were selected for but have no functions (Adams 1979, 513-514).

Faber also argues that selection as it occurs in natural selection is not sufficient to generate purpose, but for a different reason. "To qualify as purposive it must have, first, both a separation mechanism and an independently identifiable selection criterion, which, according to Wimsatt's proposal, may be called its purpose" (Faber 1984, 91). In natural selection, the separation mechanism is the differential survival of genotypes, but there is no separate criterion on the basis of which the separation is made.

> But in natural selection, far from solving some other problem or achieving some goal, the selection process *is* the problem the species faces. . . . There is no prior event or state which can be identified as the criterion or problem situation and which elicited the selection process. Therefore the selection process is not a goal-directed one. (Faber 1984, 92)

It should be noted that, however rare, it is possible to select with no criterion, as is sometimes done at the beginning of a playing-card trick when one is asked to pick a card. One might, of course, pick the one in the middle, the worn one, or the bent one, but one need not. One will, then, end up with a card with certain characteristics, such as being the bent one, but being bent was not a feature on the basis of which it was picked. Such examples, although perhaps not common, are sufficient to rule out the concept of a criterion as a part of the concept of selection. Nevertheless, it is certainly the case that selection is usually based on criteria. However, criteria are properties used in certain ways. They are not part of a disinterested description of the physical world but are devices used in ordering it. Criteria for selection are reasons for selection, and reasons seem to require minds.

As discussed earlier, Faber contrasts genuine selection with sorting, using the example of a swimming coach who selects her team by noting which of the freshmen class floats. "The coach selects, because she expects her charges to win a few swimming contests, but the pool only sorts. Nature does not select either; it does not look beyond the present scene of carnage and starvation" (1986, 116). Although he uses the contrast to argue for his feedback theory, the example is offered here because it reveals the presence of purpose in the case of selection and its absence in the case of mere sorting. However, if selection requires purpose, selection can hardly be used to analyze purpose. A. J. Bernatowicz, a botanist, writes, "*Selection*, for which some biologists expect absolution because its unfortunate implications are freely admitted, is hardly innocuous when it is described as entailing a choice by the environment" (1958, 1404) .

Woodfield argues that functions require ends or goals. "In ascribing functions to parts, one needs to assume that the organism as a whole has at least one end, otherwise the hierarchy of functions will lack a principle of generation" (1973, 40). This seems consonant with Wimsatt's including reference to purpose among the criteria of functions. Woodfield goes on, however, to say that "the suggested mechanism of natural selection that explains how evolution occurs presupposes that animals are goal-directed" (1973, 45). If the selection depends on the organisms already having goals, ends, or, in Wimsatt's language, purposes, then it is difficult to see how such goals or purposes can be explained in terms of selection.

The literal meaning of "to select" is "to choose." Standard dictionaries so define it. Wimsatt does not, at least on one occasion, seem to disagree. "For example, it remains to be shown that the effects of eliminating the worst alternatives and choosing the best differ in an on-going dynamical selection process" (1972, 64). In order for there to be an act of choosing, there must be something that chooses. Being capable of genuine choice, however, has long been regarded as a distinguishing mark of mind. A stone cannot choose where it will rest, a human being can, and regarding the immense range of items between, from amoeba to chimpanzee, our degree of our uncertainty about the ability to choose exactly matches the degree of our uncertainty about the existence of requisite mental life. It is unlikely that this is a coincidence.

The kind of selection that Wimsatt requires must be broad enough to include natural selection. This kind of selection would be, for example, cold weather killing off bison with short hair and not killing off a mutant variety with long hair. No subject does the selecting. The weather becomes colder and some animals live and some do not; that is all. That we do not really regard the weather or nature as selecting some and rejecting others is indicated by the fact that when we say, "Nature selects so and so," we feel that we could have said, "Mother Nature selects so and so," using words that clearly signal metaphorical status. Darwin used the concept of natural selection as a metaphorical extension of the selection performed by animal breeders in improving agricultural livestock.[5]

That the term "selection," as Wimsatt uses it, must have only metaphorical meaning is also indicated by the fact that we do not use "choice" in its place. Although we might say that nature selects bison with long hair, we do not say that nature chooses bison with long hair.

The point can be made another way. The kind of selection that occurs in natural selection is the kind that occurs when red apples fall and green ones do not. Although, after a time, all the red ones are on the ground and all the green ones are in the tree, we do not use this to justify talk of purpose, that, for example, the wind shook the branches for the purpose of dislodging the red apples. The difficulty Wimsatt faces is that if the selection appealed to is rich enough to explain purpose, it also presupposes mind; and if it does not presuppose mind, it is not rich enough to explain purpose.

Ruse

Michael Ruse also offers a natural selection analysis. He sees as the primary problem of teleological language how to avoid bringing in reverse causation. Future causation might, he believes, be a coherent concept, but it is objectionable because it may result in claiming causation when no cause exists. That would happen if the future cause did not materialize. "A question of some philosophical interest, therefore, is that of showing that the biologist's explanatory references to the future are not really future-causal or in any other way objectionable" (Ruse 1973[a], 176).

He first considers goal-directed behavior and adopts an analysis of it that is similar to Nagel's. Goal-directed behavior, he says, is marked by persistence in the face of obstacles and is analyzed in terms of disturbing and compensating variations. A disturbing variation is a change that, if left uncorrected, would move the system out of the goal-state. A compensating variation returns the system back to the goal-state or on the way to it. Since there is no reference here to later events causing earlier events, such an analysis, he notes, allows there to be goal-directed behavior without reverse or future causation. "A system like this is, I think, properly called "goal-directed", and it should be noted that it has not been necessary to postulate the existence of future causes, nor has it been necessary to suppose that the goal (i.e. achieving or returning to a G-state) must always be reached" (Ruse 1973[a], 178).

Ruse notes with surprising equanimity that such an analysis also fits many nonorganic systems: "It has proven impossible to distinguish between a biological phenomenon like sweating and a non-biological phenomenon like a swinging pendulum, because, questions of function apart, there is no essential difference" (1973[a], 192). Nagel thought there was. He regarded the pendulum challenge to his analysis of goal-direction serious enough to warrant introducing an independence criterion. A pendulum certainly does not swing in decreasing arcs in order to reach the lowest point and no one, not even Ruse, so describes it.

Ruse recognizes that an adequate account of teleology must include treatment of function statements and first considers Nagel's analysis. Saying that the function of chlorophyll in plants is to enable them to perform photosynthesis is equivalent to saying,

> (1) Chlorophyll is necessary for the performance of photosynthesis in plants.
> (2) Plants are goal-directed. (Ruse 1973[a], 182)

He is familiar with some of the criticisms made of Nagel's analysis and considers whether there is merit to the objection that an item having a function, y, need not be necessary for y to occur, that is, rejecting the necessary condition analysis, and comes down squarely on both sides. On one hand, he agrees with Nagel: "Nagel's reply is probably well-taken and chlorophyll is in some sense necessary" (1973[a], 183).

> Cows can feed their very young only if they have udders. . . . Naturally, this is not to deny that it is logically possible to find some other way of feeding very young cows. The point is that, as things stand at the moment (and excluding, of course, the interference of man), unless mature cows have udders, baby cows will starve. Thus, in a very real sense, the udders are necessary for the continuation of *Bos taurus*. (Ruse 1973[a], 194)

He does not specify what this sense is. Peter Achinstein remarks, "And what he does say seems to apply equally to statements thought of as being paradigm cases of non-necessary, 'accidental' statements" (1975[b], 753). Even Nagel's "All screws in Smith's car are rusty," he adds, satisfies what Ruse does say, for "as things stand at the moment (and excluding, of course, the interference of man), unless a screw is rusty it is not in Smith's car" (1975[b], 753).

However, he also adopts the opposite position and argues against the necessary condition interpretation.

> From a more general view-point, if we were to say that one thing *x* served a function *y*, it is doubtful that we would necessarily want to say that *x* was necessary for *y*. Suppose there were another thing, *x'*, which was an alternative way of getting *y*. We would still say that the function of *x* was to get or do *y* (and, similarly, the function of *x'* was to get or

do y), whilst admitting that neither x nor x' alone was necessary for y. (Ruse 1973[a], 183)

Ruse even describes Nagel's term "necessary" as unfortunate and suggests that Nagel should have said,

(1′) Plants perform photosynthesis by using chlorophyll. (1973[a], 183)

Though arguing both for and against the necessary condition analysis, in his own analysis of function statements, he does not appeal to necessary conditions.

In developing his analysis of functions, Ruse notes that a function statement must say more than merely that something does something, that is, more than 1′, for otherwise the statement,

(3) Long hair on dogs harbours fleas,

would justify saying,

(4) The function of long hair on dogs is to harbour fleas,

which, of course, it does not.

To discover what more is needed, consider a situation in which fleabites provide immunity from a certain debilitating parasite with the result that long-haired dogs live longer than short-haired dogs. We would be willing to say, "The function (or at least, one of the functions) of long hair in dogs is to harbour fleas" (Ruse 1973[a], 184). Since promoting survival and reproduction is commonly referred to as being adaptive, Ruse summarizes his view of function statements by saying that "The function of x in z is to do y" is equivalent to,

(i) z does y by using x.
(ii) y is an adaptation. (1973[a], 186)

This analysis differs from Nagel's in that something may exhibit adaptation without exhibiting self-regulation. "However, if I say of something that it enables its possessor to survive and reproduce (i.e. that it is an adaptation), I do not necessarily say anything at all about what might happen were circumstances to change in any sense, thus triggering a primary variation" (Ruse 1973[a], 186). The white coloration of an

Arctic animal is adaptive because it promotes survival in an environment of snow and ice. However, were the landscape without snow and ice, there presumably would be no compensatory variation and so no goal-directedness. Goal-direction should not be required, since "an adapted organism will probably be a goal-directed organism; but to say the first does not imply the second" (Ruse 1973[a], 187).

His deletion of reference to goal direction in his analysis of functions has the appearance of distancing himself from Nagel, but that appearance is deceiving. Presumably the goal that Nagel has in mind in his analysis of function statements is that of survival and reproduction. If so, it would do no violence to his analysis to express

(2) Plants are goal-directed,

as

(2′) Plants are directed toward survival and reproduction.

Ruse, as already noted, understands "adaptation" as referring to survival and reproduction, saying that

(ii) y is an adaptation,

is to be understood as

(ii′) y is the sort of thing which helps in survival and (particularly) reproduction. (Ruse 1973[a], 187)

Thus, when Nagel is interpreted as analyzing functional statements in terms of systems that are goal-directed toward survival and reproduction and Ruse analyzes them as talking about effects that contribute to survival and reproduction, the difference between them becomes insignificant.

Since Ruse follows Nagel in the self-regulation or disturbance and compensation analysis of goal direction, his analysis exhibits a number of the same defects. One of the indications that behavior is directed toward a goal is persistence in the face of obstacles. An indication that a rabbit is fleeing from the dog and not merely that it is moving along a certain vector is that when a tree is in its path, the rabbit swings around it and continues running, when a fence is in its path, it scrambles under and continues running, and so on. However, water flowing downhill exhibits similar behavior to obstacles placed in its path. Further, as noted in the discussion of Nagel, behavior can be directed to a goal even when there is no persistence in the face of obstacles simply by selecting cases in which there are no obstacles. Even with obstacles, the behavior may show no persistence because of physiological weakness, as in the case of starvation or weak motivation. Persistence in the face of obstacles as a test of behavior directed toward a goal is no more adequate for Ruse than it was for Nagel.

Ruse's analysis also resembles that of Canfield, who, we recall, holds,

> A function of *I* (in *S*) is to do *C means I* does *C*; and if, *ceteris paribus,*
> *C* were not done in an *S*, then the probability of that *S* surviving or
> having descendants would be smaller than the probability of an *S* in
> which *C* is done surviving or having descendants. (Canfield 1964, 292)

Ruse says that, as we have seen, "The function of x in z is to y" is equivalent to "(i) z does y by using x" and "(ii) y is an adaptation." Where Canfield, in the first part of his analysans, says, "I does C," Ruse, in his first part, says, "z does y by using x." Where Canfield, in the second part, says that if C were not done, the probability of S's surviving and having descendants would be less than if C were done, Ruse says that the thing done is an adaptation and defines "adaptation" in terms of improved survival and reproduction.

Since Ruse's analysis is much like Canfield's, the strengths and weaknesses already noted in respect to Canfield's analysis will apply to Ruse's as well, except, of course, those pertaining to Canfield's ceteris paribus clause. That clause provided, among other things, for the item to be normal. Ruse has no such provision, so his analysis avoids the problem of how to determine in a noncircular way when the subject or conditions are normal, but, like Lehman's, it is vulnerable to the nose-eyeglass counterexample. The nose does support eyeglasses and supporting eyeglasses "is the sort of thing which helps in survival and . . . reproduction" (Ruse 1973[a], 187).

Further, because Wright noticed in an earlier version of this analysis the similarity to Canfield and the vulnerability to the heart sounds counterexample, Ruse recognizes that he must address it as well.[6] He responds much as Canfield responded to the pulse example, denying that organisms with heart sounds survive and reproduce better than organisms without. "As things stand, heart sounds just seem to be a by-product of a beating heart, and thus I cannot see that an organism without the heart sounds (but with everything else) would be any less likely to survive and reproduce than an organism with heart sounds" (Ruse 1973[a], 188).

If the heart sounds challenge can be removed by imagining an organism without heart sounds but "with everything else," then legitimate function statements would be removed as well. Thus, one could argue that the function of the heart is not to circulate the blood, for a body without a heart but with everything else, including circulation of the blood, would survive as well. One could argue that the function of the stomach is not to digest food, for an organism without a stomach but with everything else, including nutrient rich blood, would survive as well. Similar arguments could be constructed against the commonly accepted function statements of fur, gills, wings, etc. Clearly, a defense that simply posits the effects and ignores how the effects are achieved goes too far.

As noted earlier, the heart sounds challenge is an example of extraneous or side effects of the item having the function. In addition, Ruse, like Canfield, has no way to exclude intermediate causal links, such as the pulse. The heart produces a pulse and a pulse contributes to survival and reproduction. Hence, on the basis of Ruse's analysis, the pulse must be a function of the heart. Presumably, Ruse would reply as he did in the heart sounds challenge and, as was seen above, inadvertently throw out all legitimate functions along with the pulse.

Cummins, in discussing Ruse, makes the striking observation that if a structure continued to be exercised in an environment so altered that it was now useless or even harmful, we, nevertheless, would continue to consider it as having the old function. "If, for some reason, flying ceased to contribute to the capacity of pigeons to maintain their species, or even undermined that capacity to some extent, we would still say that a function of the wings in pigeons is to enable them to fly" (1975, 755). This is different from the problem of vestigial organs, for vestigial organs, as the wings of ostriches, are not exercised in the new environment. Cummins' observation is also effective against Canfield and challenges any analysis that depends on evolutionary success. "Flight is a capacity that cries out for explanation in terms of anatomical functions regardless of its contribution to the capacity to maintain the species" (Cummins, 1975, 756). Nagel, in discussing Ruse, makes the same observation.

> For example, the fur of polar bears helps prevent heat loss in the animal, so that in arctic regions possession of heavy fur has an adaptive value for the animal. But what if the environment of polar bears were changed, whether because of long-lasting climatic changes in the polar regions or because of a migration of polar bears to other climes? In that eventuality, possession of heavy fur may no longer contribute to the survival and reproduction of the bears, although it might still be

> maintained that one function of the fur is the prevention of heat loss in
> those animals. (Nagel 1977, 298)

Nagel notes also that some traits that are not known to be adaptive are
still designated a function of some item. An organism's color is
considered a function of a certain gene even though the color has no
known adaptive value.

> For example, certain genes of the yellow onion produce the yellow color
> of the plant, so that the production of this color is a function of those
> genes. However, although yellow onions are resistant to a fungus disease
> while white onions are susceptible to it, the *color* of yellow onions
> appears to have no adaptive value in itself. (1977, 298)

Since color is normally a function of genes, the very many cases in which
the color has no camouflage or mating value furnish numerous examples
of this kind.

Achinstein also criticizes Ruse's requiring functions to be adaptive.

> Let us assume that there are cows of kind K that are not well-adapted,
> i.e., which do not have a good chance of surviving and reproducing. But
> let us suppose that such cows do sometimes reproduce, though in
> frequency much less than well-adapted cows, and that their offspring,
> though not very healthy, are fed from the udders of their mothers. Is
> there any reason to have to suppose that the function of udders on these
> mothers is any different from what it is on well-adapted cows? If not
> then if we can explain why the well-adapted cows have udders by
> appeal to the fact that udders serve the function of supplying food for
> their offspring, in the same manner we ought to be able to explain why
> cows of kind K have udders. (Achinstein 1975[b], 753)

In spite of the fact that Ruse adopts the disturbing variation-
compensating variation analysis of goal-directed behavior and a
doing-adaptation analysis of function statements, he takes the surprising
position that these do not support translation and that teleological
language cannot be translated into nonteleological language: "There is an
irreducible teleological element in biology" (1973[a], 195). "Hence, I
conclude that in a sense . . . since we find it illuminating to consider the
organic world with respect to its future as well as its past, biology has an
untranslatable teleological flavour distinguishing it from the physical
sciences (or at least, from most parts of the physical sciences)" (1973[a],
196).

Though regarding teleology as irreducible and its language untranslatable, he, nevertheless, also holds that teleology may be eliminable by replacing teleological explanations with nonteleological ones. "One might just replace every functional explanation with a nonteleological explanation—for example, one might explain the udders on present cows by reference to selection on *past* cows, rather than by reference to what one thinks that present-cows' udders will do" (Ruse 1973[a], 196). Adopting both of what certainly have the appearance of being incompatible positions leaves his view of the status of teleological explanation and his own analysis a mystery. Most explanations in science that have been replaced were replaced because they were regarded as being in error, as when the explanation of tuberculosis in terms of night air was replaced by explanation in terms of tubercle bacilli and the geocentric explanation of the movement of celestial bodies was replaced by a heliocentric one. Sometimes the replacements were made because the new explanation was more useful. However, if teleological explanations are held to be both not in error and useful, there seems to be no reason for replacement. There is no apparent way that teleological explanations can be irreducible and needed, but, at the same time, replaceable.

Since Ruse ties his analysis of functions to adaptation and adaptation applies only to organisms, his theory of functions does not apply to artifacts. He is aware of this. Achinstein observes, "Unless Ruse can supply some argument to the contrary, it would seem desirable to provide an analysis of functions that applies to artifacts as well as living things. If such an analysis is not possible it would be interesting to know why" (1975[b], 750).

Although Ruse endorses an analysis of goal-directed behavior and of functions as indicated above, he also regards teleological language as metaphorical. "Let me sum up now what I have tried to say or hint at so far. I argue that the teleology in modern biology is analogical. The organic world seems as if it is designed; therefore we treat it as designed" (Ruse 1982, 304). Again, "My answer is that teleology is possible because the organic world is design-like" (1982, 305). Yet again, "We get the teleology of functions in biology by analogy from human teleology, namely the teleology of human artifacts" (1982, 306), and "Teleological explanation is metaphorical, being a transference from the human conscious case. We think in terms of ends in the case of human actions and intentions, and then we translate this kind of thought metaphorically to other situations" (Ruse 1978, 199-200).[7]

In comparing one thing to another, one, by implication, rejects their being one and the same. Classifying the comparing of a fist to a hammer or a cloud to a fleece as metaphors not only claims that fist and hammer and cloud and fleece are similar, but also that a fist is not a hammer and a cloud is not a fleece. In the same way, when saying that organs can be viewed metaphorically as designed, one is also claiming that organs are not designed. Ruse expresses this point clearly when he says, "The organic world seems as if it is designed. . . . Of course, we do not today think biological phenomena really are designed" (1982, 304) and "The organic world is design-like" (1982, 305).

Although analogy may be useful in illuminating obscure features, it is not in itself an explanation why those features are there. Where analogical reasoning seems to allow inference, it is because the common property is causally, physically, or conceptually connected to the inferred property, and insofar as it is, it goes beyond analogy. Compare reasoning from the fact that ginseng root resembles human form to the folk claim that ginseng is a curative and tonic to reasoning from the fact that trees with large crowns are unstable in a gale to the claim that a sailboat with large sails will be unstable in a gale. In the first case, the inference fails; in the second, it succeeds. In the first, the common property is not causally or physically linked to the inferred effect, while in the second, it is.

When Ruse says that organisms are as if they were designed, he means that, although similar in some ways to things designed, they are not designed; that is, he denies any relevant causal relations between the appearance of design and design. However, a mere comparison between two things, with no causal connections assumed, cannot be used to explain why either has the character it has. If teleological language in biology were metaphorical, as Ruse asserts, then *it could not also be explanatory*, which he also asserts. If he is correct when he says, "I argue that the teleology in modern biology is analogical" (1982, 304), he must be wrong when he says, "My position . . . is that the teleology is fairly hardline. I would argue that when the biologist says *x* exists in order to *y*, or the function of *x* is *y* or (*y* ing), then the earlier *x* is being explained in terms of the later *y*" (1982, 305).

Ruse resolves the problem of teleology in the biological sciences by endorsing incompatible positions, both compactly embraced when he says, "My answer is that teleology is possible because the organic world is design-like" (1982, 305). If the organic world is design-like, then the only teleology possible is metaphorical teleology, making the explanations only metaphorically teleological explanations, which can only be *like* teleological explanations, and, since items metaphorically related must be distinct, not teleological explanations. If teleological explanations are not genuine explanations, then he should not have also endorsed George C. Williams, which he does, when Williams says,

> Thus I would say that reproduction and dispersal are the goals or functions of purposes of apples and that the apple is a means or mechanism by which such goals are realized by apple trees. By contrast, the apple's contributions to Newtonian inspiration and the economy of Kalamazoo County are merely fortuitous effects and of no biological interest. (Ruse 1982, 304-305; Williams 1966, 9)

Finally, if teleological language were metaphorical, there would be no need for Ruse's earlier modified Nagelian analysis, for that offered a schema in which teleological language drops out.

Neander

Karen Neander has proposed and ably defended another analysis of biological functions in terms of natural selection. The general problem in analyzing teleological explanations is how to account for the forward orientation without implying backward causation. "Teleological explanations explain the means by the ends; a development or trait is explained by reference to goals, purposes or functions, and so the explanans refers to something that is an effect of the explanandum, something that is forward in time relative to the thing explained" (Neander 1991[b], 455). The problem, she observes, is easily resolved where there is an agent having an intention, for example, a person refusing an alcoholic drink because he intends to lose weight. The intention supplies the forward orientation, yet occurs prior to the action (1991[b], 456). The parallel problem of how to account for the forward-looking character of artifacts is answered by noting that they are designed for a purpose. Since the designing occurred before the effect, the forward orientation does not upset the temporal order of causation (Neander 1991[b], 457).

Such a solution to the problem of the forward reference of functions will not suffice for biological functions, where no intentional agent is involved; yet the explanatory power of referring to functions is strong. "That the koala's pouch has the function of protecting its young does seem to explain why koalas have pouches" (Neander 1991[b], 457). The solution, she says, to the forward reference problem of biological functions, as in the case of artifact functions, lies in looking to the past rather than the future, specifically, "the simple idea that a function of a trait is the effect for which that trait was selected" (1991[b], 459). Since it might not be obvious how reference to the past can produce the reference to the future, she explains, "Selection is always of types, not tokens. . . . it is clear that the unit of selection is not a trait of an individual organism" (1991[b], 460). She summarizes her analysis of biological functions as follows:

> It is the/a proper function of an item (X) of an organism (O) to do that which items of X's type did to contribute to the inclusive fitness of O's ancestors, and which caused the genotype, of which X is the phenotypic expression, to be selected by natural selection. (1991[a], 174)

Neander repeats that selection is of types, not tokens. "Biological proper functions belong primarily to types and only secondarily to tokens because natural selection does not operate on individuals or their biological parts and processes" (1991[a], 174). The reason for its importance, as noted above, is to provide a way for a historical account to have forward reference, for she says,

> Since evolved biological functions belong principally to types, not tokens, the forward-reference to a trait's function, to what the trait is supposed to do, serves as an implicit reference to past selection of that type of trait for that type of effect. We have here, in common with other teleological explanations, an explanans that explicitly refers to something that postdates the explanandum. (1991[b], 461)

She does not mean that all selection is of types, for she lifts the type restriction for artifact functions. "Unique inventions, like the additions to James Bond's brief case, can have proper functions peculiar to them because they can be individually selected for particular effects" (Neander 1991[b], 462).

In everyday examples of selecting, as selecting red marbles from a bowl of red and blue ones or firm apples from a barrel of firm and soft ones, the selecting is done by picking up first this red marble, transferring it to another container, then picking up that red marble, etc., or grasping and perhaps transferring that firm apple from among several soft ones, then the one over there, etc. These are actions on individuals. Granted natural selection is different; still, it needs to be explained what there is about natural selection that makes what is selected ontologically different.

Daniel Hausman, in an article on the relation of explanation to causation, writes,

> I take token causation to be a relation among facts about the values of variables. . . . In nonquantitative cases such as the striking of a match causing it to light, the striking and the lighting are values of dichotomous variables. Causal generalizations, such as "The lengths of the shadows flagpoles cast depend on how tall they are", are generalizations over token causal claims, such as "The fact that the shadow of this flagpole is 30 feet long causally depends on the flagpole being 40 feet tall". . . . I will say nothing about "type causation" (and I doubt whether there is such a thing). (Hausman 1993, 437)

If causation is, at the primary level, a relation among events, as is commonly held, it is not among types or kinds, for kinds of events are not events. Claiming that causation acts on types directly, rather than acting on types via acting on individuals, has major metaphysical implications, bringing in the grand issues of the realism of universals and Platonism, and needs an extensive general defense.

It is possible, of course, to conflate types with pluralities. Neander writes, "A particular piece of genetic material, or a particular instance of a trait . . . cannot be selected by natural selection which operates over whole populations" (1991[a], 174). A population, however, differs from an individual only in number, not in ontology. Just as several items differ from an individual item by a few, a population differs from an individual by many; that is all. A population, being composed of individuals, can hardly be an abstract entity when its constituents are not. Like a pair or a dozen, populations are all on this side of the great divide; types, kinds, sets, and classes are on the other. No one questions the reality of any number of individuals and causal relations among them, as, for example, a succession of cold days causing the ground to freeze, but that is not causality among types.

Indeed, the carefully chosen words of Neander's formal analysis do not support her case. The words "It is the/a proper function of an item (X) of an organism (O)" attributes function to an item and an organism, both individuals. When describing natural selection, she writes, "which items of X's type did to contribute to the inclusive fitness of O's ancestors." The expressions "items of X's type" and "O's ancestors," referring to items and ancestors, are also about individuals, not types. Again, in observing that organs do not lose their functions when malformed or diseased, her statement, "The heart that cannot perform its proper function (because it is atrophied, clogged, congenitally malformed, or sliced in two) is still a heart" (1991[a], 180), in referring to isolated, perhaps even rare, cases, must likewise be about individuals rather than kinds.

Kinds, types, sets, and classes give mathematics and logic their atemporality. If teleology could be analyzed in language equally removed from particulars embedded in time, perhaps an analysis in terms of kinds and types could generate the forward orientation exhibited by teleological language without requiring the dreaded reverse causation. Abstract entities, however, are sterile. Just as triangles can produce stable structures, but triangularity cannot, individual hearts can beat in order to circulate the blood, but the type or kind heart cannot even beat, let alone beat in order to accomplish anything—or if one insists that linguistic usage says it can, it is only at second hand and in a manner of speaking. The ontological gulf can be compactly observed in the sharp contrast between the atemporality of types and the temporality of causation. Faber's query seems unanswerable: "How could one assign functions to a type of thing prior to discovering functions of individual tokens of the type?" (Faber 1986, 93).

When examined up close and in detail, it is difficult to see how a purely historical account, of whatever complexity or length, could justify the future orientation of teleological language. Any example will do. Suppose that, as climate slowly warms, the fur of the Arctic fox gradually changes from white to brown. Regarding a certain generation, saying that six were slightly darker than the rest and survived and reproduced at a slightly higher rate than the rest does not mean or imply (though it might suggest) that they were darker in order to improve survival and reproduction rates or that the darker fur was supposed to have that effect. If a historical account of one generation will not justify such claims, however weak or modest, by what mechanism will ten, a hundred, or a thousand repetitions of that account do so? Barbara Horan observes, "A functional explanation is 'forward-looking'. . . . An evolutionary explanation, by contrast, is historical, or 'backward looking'" (1989[a], 135). A historical account of anything carries no future reference. It may suggest such a reference, but that should prompt concern about the origins of that suggestion rather than be viewed as a discovery.

The failure of natural selection to provide future reference extends for much the same reasons to norms. The fact that functions are normative is of enormous importance. Not only are functions the effects that are supposed to occur, but parts of organisms, especially organs, are typically defined by what they are supposed to do. "For instance, 'heart' cannot be defined except by reference to the function of hearts because no description purely in terms of morphological criteria could demarcate hearts from non-hearts" (Neander 1991[a], 180). Further, the concept of malfunction, which is critical for descriptions of much behavior, is based on norms. Yet, what is supposed to occur or ought to be cannot be logical derived from what is, or, rather, what was, and a historical account describes only what was.

Neander makes no provision for the large portion of teleological language used to describe goal-directed behavior of lower organisms, something usually regarded as distinct from functions.

Summary

The many difficulties in devising a naturalistic understanding of teleology seem to arise in one way or another from its future orientation, its appeal to the later to explain the earlier. This is more apparent in the case of goal-directed behavior than in functions, for the goal is realized, if ever, after the behavior occurs. If reference to the future is real, this does pose a significant problem for any natural selection analysis, since that analysis appeals not to the future, but to the past. It is significant that the five analyses just examined appeal to natural selection only in respect to functions, not in respect to goal-directed behavior. Canfield, Lehman, Wimsatt, and Neander do so explicitly and either completely ignore goal-directed behavior or mention it only in passing. Although Ruse discusses goal-directed behavior, the analysis used for that part is not natural selection but rather a variant of Nagel's disturbing variation-compensating variation analysis. If a historical account of what happens during one generation of fruit fly does not generate future reference, two, several, or many will not. Changing the time order of either causation or of reference is not accomplished by multiplying instances. Some believe one can get the needed future orientation by construing natural selection as acting on types rather than individuals, but the details of how this is accomplished are always left vague. It is never explained how causation works among abstract entities, which, although they may be about things temporal, are, themselves, atemporal. Even if type causation were accepted (other than, of course, as a shorthand way of referring to causation among particulars), it is never explained how moving from tokens to types produces time reversal.

Nagel, Cummins, Achinstein, and Faber raise serious questions about whether functions depend at all on improvement in survival and reproduction. If organs and parts, such as fur, wings, udders, cilia, and some color-producing genes have functions even when they do not contribute to improved survival and reproduction and may even lower it, a natural selection analysis must confront significant exceptions. Paul Griffiths argues, "Traits arise and spread for non-adaptive reasons, perhaps as side-effects of adaptive traits, and only later acquire a function. It is also common for traits to lose old functions and acquire new ones" (1993, 414).

A natural selection analysis seems unable to separate effects that are functions, such as circulation of the blood, from those that are side effects, such as heart sounds and the production of lactic acid and carbon dioxide. One can, of course, impose a theory and say that selection tracks causation and that what caused survival was circulation, not heart sounds, but the dying and surviving show nothing of that. Such an analysis also seems unable to separate functions, such as circulation of the blood, from intermediate effects, such as the pulse and moving blood through the component vessels. If organisms must be organized the way they are, then not only function effects, but also side and intermediate effects, are necessary. If organisms may be different from the way they are, side and intermediate effects are not necessary, but then neither are the effects that are the functions.

The traits acted upon by natural selection must be heritable traits. That requires the mechanisms of heredity to be already in place for natural selection to operate. This blocks appealing to natural selection to explain the functions of those mechanisms. If protomechanisms of heredity are proposed, the same question can be asked about them. The parts and processes of such protomechanisms must themselves have had earlier functions that enabled an earlier stage of natural selection to operate; hence, that earlier stage of natural selection cannot be used to explain those earlier functions.

Several have commented on the fact that teleological language has undergone no change in meaning over the centuries. Frankfurt and Poole report, "The concept of biological function is an old one, and numerous standard descriptions of functions do not appear to have changed in meaning with the advent of evolutionary theory" (1966, 71). Ruse agrees: "When I read the *Origin*, I am struck by the extent to which the teleology changes from previous non-evolutionary writings. 'But the teleology does not change!' 'Precisely!'" (1986, 60) Elizabeth Prior observes, "Physiology is a lot older than evolutionary biology and physiologists were making claims about the functions of organs long before there were any claims about their evolution. For example, in 1628 William Harvey was able to claim that the function of the heart is to pump blood" (1985, 317). Canfield, in an essay laying out an entirely new analysis, says, "But to read 'function' in terms of survival and having progeny is to read back into an already extant method of biological investigation ideas that belong properly to post-Darwinian times. I no longer think it is plausible to assume that the meaning of 'function' has changed *tout court* with the advent of evolutionary theory" (1990, 42). When we read a sentence such as "The function of quills is to protect from predators," we do not first check the date when it was written. Shakespeare, in *A Midsummer Night's Dream*, has Hermia say, "Dark night, that from the eye his function takes, the ear more quick of apprehension makes," and modern audiences understand readily. The fact that when we read teleological expressions, including those ascribing functions, we understand them without adjusting for date indicates that teleological language has been stable. Had the meaning changed, lexicographers would have long since documented it.

There is also the problem natural selection analyses face with Sorabji's luxury functions, discussed earlier. There seem to be such functions, functions that do not contribute to survival but to other ends, such as pain relief. If there were innate faculties responsible for sensitivity to beauty, it would be hard to see that reducible to survival.

If it turns out that some organisms have a biological clock governing aging and time of death, that also will be difficult to understand in terms of improved species survival. Arguments for functions referring to species survival normally appeal to the fact that the feature decreases the probability of death of the parent or the offspring. Since death is the very opposite of survival, a part or process that controls aging and death should not, on pain of incoherence, be construed as a means for survival, even for lemmings and salmon.

If selection in natural selection is a metaphor, then the explanatory effectiveness of teleological statements is undercut. One may describe clouds metaphorically as fleeces, but that is not of much value if one is trying to explain why it rains. In order for natural selection to explain, the selection must be genuine. However, genuine selection requires a selector, something capable of considering alternatives rather than something capable only of doing something that resembles selecting in some way or other. Wind and rain, heat and cold, climatic and geological changes do not suffice, and personifying nature, such as giving it foresight and wisdom or parenting qualities, brings us back to metaphor.

If history determines function, knowledge of function should be grounded on knowledge of history, for inferring history from function would make the thesis trivial. However, we have greater confidence in our knowledge of an item's function than in our knowledge of its history. This is due in part to the required inference being across vast periods of time. Amundson and Lauder note, "Many structures are ancient, having arisen hundreds of millions of years ago. During this time, environments and selection pressures have changed enormously. How are we to reconstruct the ancient selected effect?" (1994, 461). Linking function to selection is also difficult because selection involves changes in supporting structures. "Thus, selection for increased running endurance in a population of lizards may have the concomitant effect of increasing heart mass, muscle enzyme concentations, body size, and the number of eggs laid, despite the fact that selection was directed only at endurance" (1994, 461). The result is that functions are delineated more sharply than natural selection can select. "If we cannot identify the causal relationships among these correlated variables to single out the one that was selected for, we will be unable to assign a trait X to the SE function already identified" (1994, 462). Further, the thesis that history determines function is uninformative if the history must be inferred from the function.

Notes

1. See also Nissen (1970, 193-95).

2. Adams (1979, 511) makes a similar observation, but uses an example about building construction even though Lehman limits his analysis to biology.

3. It makes little difference whether one says, to use a simple analysis, "'The function of X is Y' can be translated into 'X is a necessary condition for Y'" or "The function of X can be identified by finding that for which it is a necessary condition." Wimsatt's rejection of a translation analysis might be connected with his rejection of appealing to ordinary language. "It may be simply that many of the logical features of teleology and purpose are found far beyond the range where we are willing (due to 'ordinary language' prejudices) to apply them" (1972, 16). "Despite the protestations of some 'ordinary language' philosophers, our ordinary conceptual scheme for the explanation of human intentional action is open to revision and is constantly being revised" (1972, 65). Inserting a disclaimer appears intended to block traditional examples of teleological items being used in criticism. "I have attempted to give an extensive and

detailed survey of the logical structure and implications of the use and meaning of function statements. The analysis proposed has been partially, but not I think implausibly, reconstructive and normative" (1972, 80). The result is that all criticism, other than that of logical inconsistency, is barred.

4. Later Wimsatt (p. 40) furnishes a somewhat more detailed version, "$F(B(i)_{P,T}, S_{P,T}, E_{P,T}, P, T) = C$," which shows the restraints supplied by P and T on what values B, S, and E take.

5. However, Robert M. Young (1971, 442-503) makes a very strong case that Darwin's language was not clearly metaphorical but was, rather, ambiguous between metaphorical and literal usage. Regarding Darwin's talk of selection, he says (p. 455), "In moving from artificial to natural, Darwin retains the anthropomorphic conception of *selection*, with all its voluntarist overtones."

> Anthropomorphic, voluntarist descriptions of natural selection occur throughout *On the Origin of Species.* . . . It will help to sharpen our sense of how remarkable this is if it is recalled that the rules of scientific explanation which were developed in the seventeenth century had banished purposes, intentions, and anthropomorphic expressions from scientific explanations. (1971, 461-62)

Young notes that Darwin repeatedly speaks of natural selection as a power, as something that scrutinizes, intently watches, picks out with skill, has a motive, etc. When criticized for such language, Darwin faulted his critics for misunderstanding him, even though such language was routinely used in describing domestic breeding programs and the criticism arose because in such contexts the action of selecting has an agent capable of performing the selecting while natural selection does not. He claimed to use anthropomorphic language for reasons of brevity, but neutral terms, such as Spencer's "survival of the fittest" were as brief.

6. Wright discusses this (1972[a], 512-514) based on Ruse's earlier version (1971, 87-95).

7. Ruse claims agreement with Wright regarding metaphor in teleology, but Ruse says teleological statements are metaphorical and Wright says they evolved metaphorically but are now dead metaphors. A dead metaphor is no longer a metaphor.

Chapter 4

Taylor

Charles Taylor's analysis of teleology appears as a subsidiary theme in *The Explanation of Behaviour*, a book devoted primarily to a critical appraisal of behaviorism in psychology. It is included because it attracted considerable attention when it appeared and because Wright viewed it as a precursor to his own analysis. He defends explanation containing reference to purpose. Purpose is connected to teleological language in that behavior that is purposive is behavior performed for the sake of something, and such behavior can be described teleologically. "For to explain by purpose is to explain by the goal or result aimed at, 'for the sake of' which the event is said to occur. Explanation which invokes the goal for the sake of which the explicandum occurs is generally called teleological explanation" (C. Taylor 1964, 5-6).

Purpose is not an entity that causes anything. It is not something internal to the organism. It "cannot be identified as a special entity which directs the behavior from within" (C. Taylor 1964, 10), but is, rather, a publicly observable feature of that behavior.

> The claim is that animate beings are special in that the order visible in their behavior must itself enter into an explanation of how this order comes about. This can in part be expressed by the claim that the events which bring about or constitute this order are to be accounted for in terms of final causes, as occurring "for the sake of" the order which ensues. Now this claim is not inherently "mystical" or non-empirical in nature . . . nor does it entail postulating any unobservable entities. (C. Taylor 1964, 17)

Taylor's interest in teleology centers on the question of whether a teleological law can be replaced by a nonteleological law with no explanatory loss. His illuminating explorations have prompted much commentary. The context of the use of a teleological law, he believes, conveys the claim that teleological laws cannot be reduced to anything else. "For the claim that a system is purposive is a claim about the laws holding at the most basic level of explanation" (C. Taylor 1964, 18). Again, "Now it is clear that the claim that the behaviour of a system must be accounted for in terms of purpose or 'natural' or 'inherent' tendencies concerns the laws which hold at the most basic level of explanation" (1964, 20), adding, "Thus the claim that 'the purposes' of a system are of such and such a kind affects the laws which hold at the most basic level. In other words, it is incompatible with the view that the natural tendency towards a certain condition can itself be accounted for by other laws" (1964, 21). Taylor sees teleological laws as setting the conditions of what needs explaining and what does not, somewhat as the Aristotelian view did by talk of natural and violent motions. Normal behavior of an animal is accounted for by teleological laws, while abnormal behavior is explained by nonteleological laws referring to antecedent conditions, such as interfering factors. However, he also holds that whether teleological laws are really basic is both a conceptual and an empirical question. "Rather it is the type of issue between what Thomas Kuhn has called different 'paradigms.' This is ultimately decided by which better fits the facts, by which generates soluble problems, and is thus empirical" (C. Taylor 1968, 124). Most of the ample discussion of this book has not been about his analysis of teleology but about the status of teleological language, especially the status of teleological laws and whether they are compatible with nonteleological causal laws. This issue will not be addressed here, but only the usual question whether his analysis of teleological language is plausible.

In explaining the open and public character of teleological explanation, Taylor gives, in effect, if not by intention, three distinct accounts. One account is that a teleological description of the form "B occurs for the sake of G" means "B brings about G." "Now when we say that an event occurs for the sake of an end, we are saying that it occurs because it is the type of event which brings about this end" (1964, 9). Such a position would resolve the problem of relating teleological and ordinary causal description by denying any difference. However, saying, "Rainfall brings about the growing of grass" or its equivalent "Rainfall causes the grass to grow" is not the same as saying "Rain falls in order for the grass to grow." The world is full of events, as rocks falling, vapors rising, and winds blowing, which are described causally but not teleologically.

Further, implying uniform success, as the above does, makes no allowance for behavior ending in failure.

A second position discernible says that "B occurs for the sake of G" means "B tends to bring about G." "Thus an explanation of a teleological type does involve the assumption that the system concerned 'naturally' or inherently tends towards a certain result, condition or end" (C. Taylor 1964, 17). The problem of not distinguishing teleological descriptions from ordinary descriptions remains, for saying, "Rainfall tends to make the grass grow" is not the same as saying, "Rain falls in order for the grass to grow." Counterexamples are easily generated based on incorrectly equating probabilistic causal statements with teleological statements. Saying that B tends to bring about G is, however, an improvement over saying that it brings about G because allowance is made for occasional failure. This change does not, however, provide for the fact that some behavior is for the sake of things it does not even tend to bring about, thus does not provide for behavior directed toward impossible goals.

Although Taylor makes the statements just discussed, he probably did not intend them to be taken as fully endorsed positions. The position he clearly endorses and the one with which he has become identified is a third one, that "B occurs for the sake of G" means "B is required for G." "To offer a teleological explanation of some event or class of events, e.g., the behavior of some being, is, then, to account for it by laws in terms of which an event's occurring is held to be dependent on that event's being required for some end" (C. Taylor 1964, 9). Being required suffices: "For what is claimed by the teleological explanation is that what occurs is a function of what is required for the system's end, *G*, that an event's being required for *G* is a sufficient condition of its happening" (1964, 15).[1] Again, "to say that anything is a goal for animals of a given species is to say that they will do whatever is necessary *within the limits set by their motor capacities* to attain it" (1964, 223)[2] and "An explanation is teleological if the events to be explained are accounted for in this way: . . . *B* is explained by the fact that *S* was such that it required *B* for *G* to come about" (1970, 55).

He sees the teleological account of behavior as different from a causal account and considers it to be an empirical matter whether a teleological or a causal explanation is more appropriate. There has been considerable discussion of his account of the relation between the requirement formula and causal explanation and of his suggestion that teleological explanation, understood in terms of his requirement analysis, is a basic level of explanation (C. Taylor 1964, 18-25).[3] The position becomes somewhat clearer in his response to Denis Noble.

> Suppose it be the case that the set of correlations (SE)1→B1, (SE)2→B2......(SE)n→Bn exhibit no intrinsic order, so that they leave us just as incapable of predicting what will happen in situation (SE)n+1 as we would have been before establishing this set of correlations; suppose in other words it be the case that these correlations are not instances of a general non-teleological law, B=f(SE); then we would clearly have a teleological account without a corresponding non-teleological one. For the correlations T1→B1......Tn→Bn do permit extrapolation to new cases: we know for each new situation that that B will occur which is required by this situation if G is to come about. We have only to examine situation n+1 to see what it requires for G and we can predict what B will be like in this situation. (C. Taylor 1966-67, 142-143)

Even if it were true that every individual case describable teleologically could also be described nonteleologically, teleological descriptions would be distinctly different from nonteleological ones because one could know the teleologically described situation without knowing the nonteleologically described situation. On the basis of this passage, it appears, as Douglas Porpora notes, that Taylor intends teleology descriptions to apply not to an individual instance of a behavior but only to a range or class of behaviors occurring under varying circumstances (Porpora 1980, 571).

Among the virtues of this analysis, Wright, who sees it as a precursor of his own, says that it accounts for the forward-orientation of teleological statements, provides the needed contrast between teleological and ordinary causal explanations, does not require knowledge of internal mechanism, and allows determination by direct observation (Wright 1976[b], 34-35). In addition, it provides for goal-failure. Nevertheless, it seems to be an analysis one can hold only if one believes counterexamples do not matter. After all, Nagel, who wrote several years earlier, began his analysis with a requirement or necessary condition version, saw that it would not work, and supplemented it with the self-regulatory system account.

Although Taylor's requirement analysis of teleology seems intended to provide for the plasticity of goal-directed behavior, it fits best the stereotyped behavior of simpler organisms, for if an organism is capable of only one behavior, then that also is the one requiring least effort in its repertoire.

Taylor sees little difference between saying B brings about G and saying B is required for G.

> This means that the condition of the event's occurring is that a state of affairs obtain such that it will bring about the end in question, or such that this event is required to bring about that end. . . . The difference is small between these two formulae. To say that an event is required for an end is to say more than that it will bring it about; for it adds that no other event, or none other in the system concerned can bring it about. (C. Taylor 1964, 9)

On the contrary, to say that an event is required cannot very well say more than that it will bring it about inasmuch as it does not say at all that it will bring it about. To say that an event is required for something is just different from saying that the event will bring it about, and that difference is not small. It is all the difference that separates necessary conditions from sufficient conditions. Taylor's unusual position is not a passing fancy, for he repeats it several years later (1970, 55).

Since Taylor's analysis explains behavior in terms of what is in fact required rather than what is perceived as required, it does not fit the large class of goal-directed behavior based on faulty perception. Counterexamples can be constructed around behavior directed to goals that are achievable but not by the behavior in question. Some consult horoscopes in order to learn their character and personality traits. Although such traits can be discovered, consulting horoscopes will not do it. Thus, although it is correct to say, "He consulted his horoscope in order to discover his character traits," it is incorrect to say, "Consulting his horoscope was required for discovering his character traits." In a maze of appropriate design, it is correct to say of a rat that it turned left in order to reach the food but incorrect to say that its turning left was required for reaching the food. N. J. Block notes that Taylor's requirement formula "suffers from the serious deficiency that it fails to take account of the representations of information about the world in organisms whose behaviour is dependent on such information" (Block 1971, 110).

Another class of counterexamples is based on behavior directed to impossible goals. Thus, it is correct to say, "The Michaelson-Morley experiment was done in order to detect ether," but it is incorrect to say, "The Michaelson-Morley experiment was required to detect ether" because detecting ether was impossible. Further examples can be constructed about traveling faster than light and building perpetual motion machines, as well as the quite ordinary ones of lifting an excessive weight or jumping a great height. Nonhuman examples can be constructed about animals searching for food that is not there, digging to escape when escape is impossible, and so on.

A third class of counterexamples is based on the fact that behavior may be correctly directed toward an achievable goal but is not required because of alternative ways to reach the goal. In his examination of Taylor, Wright observes,

> As it stands, the "because it is required" formula is at best a sufficient condition for behavior to be teleological; it is very far from necessary. Nothing a predator does by way of stalking is in any strong sense *required* for the goal of food, although the activity is clearly purposive. Prey may be trapped or found dead; or it might accidentally walk into a predator's unsuspecting clutches; or some fairly unsophisticated trickery might result in the same end without stalking. (Wright 1976[b], 35)

Aside from his concession that the requirement is a sufficient condition for teleological behavior, Wright is undoubtedly correct. Indeed, teleological behavior could not exhibit its much regarded plasticity, a feature some regard as so central as to be distinctive of teleological behavior, if alternatives were unavailable.

Oddly, Taylor seems to recognize that behavior could be directed toward a goal although it is not required for the goal, for he acknowledges, "If there are several possibilities we cannot account for the selection between them unless we add another teleological principle, e.g., of least effort" (1964, 9). He does not, however, develop his suggestion to see where it leads, or notice that, as it stands, this concession threatens his requirement analysis.

Incorporating the least effort principle would produce as a revised version of the analysis: "B is done for the sake of G" means "B is required for G or involves the least effort of those behaviors which are disjunctively required for G." Once again, counterexamples abound, such as animals taking the longer or more difficult route rather than the shorter and humans not being optimally efficient, and so on.

Taylor adds that the various behaviors considered must be within the repertoire of the subject (1964, 9). This provision must not be interpreted as including what the organism perceives as being within its abilities or as requiring least effort, for that would be incompatible with Taylor's claim that everything needed to explain behavior is publicly observable.

Even where there are alternative actions in the repertoire, it is not the case, as mentioned above, that the animal always performs the action involving the least effort. This is evident from observing the erratic flight paths of insects and birds and the roundabout routes and interrupted movement of most animals, even when the goal, such as food, is near and visible. Explaining why an easier route is avoided, for example, because of sensed danger, habit, pack hierarchy, or learned aversion, involves going outside Taylor's requirement plus least effort formula to explain.

In stating his teleological law, abbreviated "T-B," that a certain behavior's being required for G is sufficient for that behavior to occur, Taylor explains that he means the law to hold invariably, but then adds, "By 'invariably' is meant here not necessarily every time, but such that the exceptions can be cogently explained by interfering factors" (1964, 14). That would seem to take care of many of the exceptions, but Wright finds the interfering rule of little value.

> For instance, our favorite predator (suppose it's a feline), while resting between stalks, might be licking his paw, claws exposed, when a low-flying bird collides with the claws, killing itself and providing a meal for our still hungry friend. A very good case indeed can be made for saying that this rather disinterested exposure of the claws was precisely what was required to snare the unwary fowl. And if *anything* can count as interference, from having other goals to becoming tired, then it would appear to be impossible to adduce evidence against the claim that this disinterested claw exposure happens every time it is required to capture birds, barring interference of course. Hence, Taylor's formula forces us to say, what is palpably false, that the predator was licking his paw in order to catch (or, for the sake of catching) the bird. (Wright 1972[b], 209)

Thus, the formula is too indiscriminating because it includes lucky accidents. An ornithologist accidentally spotting a rare bird while jogging or James Marshall's discovery of gold while building a sawmill for John Sutter, would, on this account, justify saying the first jogged in order to spot the bird and the second built a sawmill in order to find gold.

Taylor says that "what is claimed by the teleological explanation is that what occurs is a function of what is required for the system's end, G, that an event's being required for G is a sufficient condition of its happening" (1964, 15). Wright notes that sleep is required for achieving all sorts of goals.

Whenever an animal finds itself so tired that sleep is necessary for it to continue its everyday, goal-achieving activities, it sleeps. In other words, the requirement of sleep for these goals is (qualifiedly) sufficient to produce it. So on Taylor's analysis, we would have to say that the animal went to sleep in order to achieve all sorts of goals the next day. But this is absurd. (Wright 1972[b], 209-210)[4]

Wright also calls attention to a kind of multiple goals problem. He imagines a predator that is both hungry and thirsty, sees no water, but does see what appears to be prey, and begins stalking behavior. While doing so, he comes upon water and drinks.

At this point Taylor's formula would require that we say the predator's behavior all along had been directed toward (i.e., for the sake of) obtaining water. . . . But this conflicts directly with the fact that the predator's behavior was obviously stalking; it was obviously *for the sake of* catching its prey. (Wright 1972[b], 209)

N. S. Sutherland observes, "There is no known system of which it is true to say 'an event's being required for *G* is a sufficient condition of its happening' where *G* is the goal of the system," goes on to quote Taylor, "in short the lion will charge from a point where, all things considered, the prey can best be seized," and dryly adds, "thus implying that lions are omniscient" (Sutherland 1970, 115).

Behavior that is required is required to reach a certain goal. Taylor gives little information about how that goal is determined, but does state, "To say that anything is a goal for animals of a given species is to say that they will do whatever is necessary *within the limits set by their motor capacities* to attain it" (C. Taylor 1964, 223). Seeing a single instance of apparently goal-directed behavior, for example, a dog apparently chasing a rabbit, would not suffice, for it would have to be supplemented with a great number of additional trials involving an immense variety of obstacles to see if the organism does, indeed, do whatever is necessary to reach G. This would make real-life judgments of goal-directed behavior unwarranted. Indeed, since the range of possible obstacles has no obvious limit, it appears the test could never be completed, implying that we would never be justified in a judgment about goal-directed behavior. In addition, there is an element of circularity involved, for the property of being required is used to identify both behavior directed to a goal and the goal, the difference lying only in the number of behavior sequences involved.

If G cannot be identified independently from the behavior, then other problems arise, for any piece of behavior is required for an indefinite, perhaps infinite, number of termini. A teleological description might be "The lion wades the stream in order to catch the antelope" and would be analyzed as "Wading the stream is required for the lion to catch the antelope." This information does not enable one to know that catching the antelope is the goal. After all, wading the stream is also required to reach the shade of the acacia tree on the far bank and to stalk wildebeests in the valley beyond, as well as all sorts of irrelevant things, such as causing turbulence in the stream, casting a lion-shaped shadow on the far bank, and reducing the distance to London. Robert Borger, in commenting on Taylor's analysis, observes, "All situations have unlimited potential for developing into others, and there is nothing at this stage of the formulation which accounts for the selection of any particular facet of this potential" (Borger 1970, 83).

If the goal were to be read off behavior, the behavior of inanimate objects would not even be excluded. Seasonal increase in solar heating is required to melt the ice on the Mississippi River; yet we do not say that the goal of such increased heating is to melt the ice or that solar heat increases in order to melt the ice. A certain precise positioning of the moon between the earth and the sun is required for a sun eclipse visible in Cardiff, but the eclipse is not regarded as a goal of such positioning. Examples can be generated endlessly of this being required for that, such as moving electrons being required for magnetic fields, the considerable density of rocks being required for rocks to sink in water and the slight density of smoke for it to rise in air, where teleological language has no place. Borger observes, "Yet 'the condition of the ground was such that it required a lot of water to turn it into mud' is hardly satisfactory as an explanation of heavy rainfall" (Borger 1970, 83).

A fourth position regarding teleology can be extracted from Taylor's investigation into the relation of intention to action. It applies only to goal-directed behavior of man and higher animals. It is limited to organisms capable of actions, with actions understood as movements governed by intentions. "Now the notion of an action as directed behaviour involves that of an intentional description" (C. Taylor 1964, 58). Again, "Thus 'action' is a notion involving that of intentionality, and the types of system to which action can be attributed are those to which consciousness or intentionality can be attributed, beings of which we can say that things have a certain nature or description 'for' them" (C. Taylor 1964, 58). He elaborates,

Thus to say that it is one thing and not another which moves a man to action is to say that his action really is directed one way and not in another, and an account of what this means involves the notion of intentionality; for the goal concerned has an intentional description for him, as the goal of his action. (C. Taylor 1964, 60)

The inclusion of intentionality changes the form of the explanation, for at issue now is not whether B is actually required for G but rather whether the subject perceives B as required.

In an ordinary teleological system with goal G, B will occur in the condition where B is required for G; (let us call this condition 'T'). But in an 'intentional system' with goal G, T is not sufficient for B. In this latter case, however, B will occur when T is *seen* to hold by the 'system' (in the absence of deterring factors), for otherwise we could not ascribe goal G to it. (C. Taylor 1964, 62)

He adds, "The situation as it really is may differ from the situation under its intentional description for the agent, that is, the intentional description may not in fact hold of it" (C. Taylor 1964, 62).

Thus, "B occurs for the sake of G" now means "B is believed to be required for G." This analysis in terms of intention is not a modification of the requirement analysis, but an entirely different analysis.[5] Even so, it will not do. Much human intentional action is not based on what is perceived as required, even if supplemented with the principle of perceived least effort. Human motivation is far more complex than such a simple two-component system would have it.

Taylor does not make clear how his analysis in terms of intentionality is to relate to his analysis in terms of behavior required for G. He calls the former "explanation by purpose" (1964, 63) and contrasts that with "ordinary teleological explanation," that is, explanation in terms of B's being required for G, suggesting that explanation by purpose is a variety of teleological explanation. However, earlier he said, "The element of 'purposiveness' in a given system . . . consists rather in the fact that in beings with a purpose an event's being required for a given end is a sufficient condition of its occurrence" (1964, 10), which removes intentionality from explanation by purpose by replacing it with necessary conditions and sufficient conditions.

Taylor is surely reasonable in relating purposive behavior to intentionality. However, it seems impossible for such an analysis to be fitted into a nonintentional analysis, such as his requirement analysis.

What Taylor alludes to as a mere modification of his overall analysis is, instead, a radically different and competing approach. If purpose involves intention and if intention cannot be reduced to publicly testable empirical laws, this modification is, in fact, incompatible with his earlier position that purpose contains nothing hidden.

Taylor does not address the large subject of functions at all.

Notes

1. Jon Ringen (1976, 230) believes Taylor's analysis can be expressed as "O has G_L as a goal if and only if O is *disposed* to exhibit behavioral plasticity with respect to G_L," making it similar to that of Braithwaite. This interpretation is hard to reconcile with Taylor's requirement formula.

2. Woodfield interprets Taylor as claiming that B is a means to G. "Taylor's main idea, that to explain B teleologically is to assert, relying on the support of a law-like generalisation, that B occurred *because it was a means to G*" (1976, 81ff).

3. See also Denis Noble (1966-67, 96-103), Taylor's reply (1966-67, 141-43), Noble's reply, (1967-68, 62-63), Timothy L. S. Sprigge (1971, 149-70), and Alan Montefiore (1971, 171-92).

4. Wright uses this example to argue that Taylor's analysis does not discriminate between functions and goals.

5. George Sher notes (1975, 32), "If we do accept it, we will have to abandon the idea, suggested by many of his phrases and hitherto taken for granted, that Taylor intends that purpose be given a *conjunctive* analysis, in terms of teleological explicability on the one hand and some further distinguishing feature on the other. For if the determining condition for *B* is that *B* is *seen to be* required for *G*, then it is evidently *not* a determining condition that *B* actually *is* required for *G*; and so explicability in terms of ordinary teleological laws will turn out to play *no* part in the analysis of purpose."

Chapter 5

Wright

By means of several articles and the book *Teleological Explanations: An Etiological Analysis of Goals and Functions*, Larry Wright has made significant contributions to the defense of teleological language in the life sciences. He analyzes teleological language both in respect to goal-directed behavior and in respect to functions in a manner that, he believes, does not presuppose mind. He offers a comprehensive analysis, one covering teleological descriptions of both human behavior and nonhuman behavior. He has many useful and important things to say about teleology. Among these is his reminder that many teleological judgments are as reliable and made with as great intersubjective agreement as are most observation judgments. No one else has made this point so strongly or expressed it so persuasively.

> Goal-directedness is often obvious on its face. Many of our teleological judgments are as reliable and intersubjective as the run of normal perceptual judgments. Occasionally there simply is no question about it: the rabbit is fleeing, the cat stalking, the squirrel building a nest. Certain complex behavior patterns seem to demand teleological characterization. (Wright 1976[b], 23)

In another place, he says, "But it is *also* clear that in some cases, the goal-directedness of behavior is obvious, palpable, instantly recognizable, and unmistakable to the normally sighted individual, which is just to say it is observable" (Wright 1976[b], 54).

From the fact that teleological judgments are often obvious and reliable and exhibit strong agreement among observers, certain consequences follow that restrict possible teleological analyses.

> Accordingly, we should view with suspicion any analysis that contends
> that goal-directedness consists in a relationship among parameters of
> which we are usually quite ignorant in the contexts of these reliable
> judgments. And in these cases we simply do not know the laws and state
> descriptions, the causal chain and variancy, the underlying mechanism.
> (Wright 1976[b], 30-31)

Such considerations immediately render negative-feedback theories
implausible. One might add that if many teleological judgments are
obvious and reliable and show general agreement among independent
observers, and if people of little training, experience, or intelligence can
make teleological judgments with great intersubjective agreement, then
what is discriminated can hardly be highly complex. Wright is surely
correct in decrying what he calls "the incredible Rube Goldberg
complexity found in much of the empiricist literature on teleology"
(1976[b], 25). It is hard to overestimate the importance of this
observation, one that seems, unfortunately, to have gone unnoticed.

Wright characterizes his analysis as a "consequence-etiological"
analysis. "When we say that teleological etiologies are consequence-
etiologies, we are saying that the consequences of goal-directed behavior
are involved in its own etiology: such behavior occurs *because* it has
certain consequences" (1976[b], 56).

If the consequences of the behavior determine that behavior and those
consequences occur after the behavior, then later events appear to
determine earlier events. Wright, nevertheless, emphatically rejects reverse
causation. "The reversal of cause and effect has long been hung like an
albatross around the neck of teleology, but it is not clear how this view
could have survived even modest scrutiny" (1976[b], 10).

The traditional way to eliminate reverse causation has been to bring in
intentions, purposes, and goals. Braithwaite describes this well: "Now
there is one type of teleological explanation in which the reference to the
future presents no difficulty, namely, explanations of an intentional human
action in terms of a goal to the attainment of which the action is a
means" (Braithwaite 1964, 324).

The difficulty with getting rid of reverse causation by bringing in
intentions, purposes, or goals is, of course, that teleological statements are
also made about subjects believed incapable of having intentions,
purposes, or goals. Using an intricate argument, Wright claims that
teleological language is extended to such subjects through a three-stage
process involving, first, teleological usage that includes reference to

intentions and goals; second, selective metaphorical extension; and, third, dead metaphors.

The first stage of the argument says:

> When I explain my going to the store by saying I went "in order to get some bread," I do not imply that the actual act of purchase caused my going which preceded it. The purchase of bread was a *goal* of the action, not a cause. Perhaps my *having* of that particular goal could be viewed as a cause of the action; but that of course is something that preceded the action, and hence is not guilty of the egregious time-reversal imputed teleological accounts of behavior. (Wright 1976[b], 10-11)

This takes care of the problem of reverse causation at the human level along the traditional lines described by Braithwaite. Although purchasing the bread did occur after my going to the store, it was a goal of my going and not a cause of it. Having the goal to buy bread might well have been a cause, but having the goal to buy bread is not the same as buying bread. Having the goal occurred before or concurrently with going to the store, so, it could well have been a cause of the action without involving reverse causation.

In the second stage of the argument, that of providing for teleological language for descriptions of nonhuman behavior, Wright extends teleological language by claiming a selective transfer of reference to properties by means of the mechanism of metaphor.

> In general, I will argue that the feature of human teleology which transfers to nonhuman cases is the fact that when we say "A in order that B," the relationship between A and B plays a role in bringing about A. It is this which is being pointed out, rather than intelligence and conscious purpose. In fact, just as in many of the nonteleological cases we have examined, it is usually so bizarre to suppose conscious intent in the nonhuman cases that nobody is seriously misled. . . . At the same time, since there is clearly no reversal of cause and effect in human cases, this argument in terms of metaphorical extension should have no tendency whatever to license such reversal in nonhuman cases. (Wright 1976[b], 21)

The argument so far is that teleological language, having its origin in descriptions of human behavior, is extended to nonhuman cases by the mechanism of metaphor. The metaphor transmits only the claim that the behavior occurred because of the consequences, that is, the claim of consequence-etiology. Thus, teleological language can legitimately be used to describe the behavior of nonhuman organisms because the part that transfers is nonmentalistic. Teleological language can be used to describe the behavior of organisms that are not assumed to have intentions, purposes, or goals because the metaphor does not transmit reference to intentions, purposes, or goals.

The third and final stage of Wright's argument explaining teleological descriptions of nonhuman behavior is that the metaphor has become a dead metaphor. There are, of course, no limits to the legitimate use of teleological expressions when used purely metaphorically. Not only the behavior of lower organisms, but also the inanimate behavior of rock slides, lightning bolts, and receding galaxies can be properly described in teleological language if that language is used purely metaphorically. The only interesting question regarding the teleological description of the behavior of lower organisms is whether such descriptions can be used literally. Wright, aware of this, provides for literal teleological descriptions of nonhuman subjects by claiming the metaphor has become a dead metaphor. "It will be the central contention of this essay that teleological expressions in most nonhuman applications represent dead anthropomorphic metaphors" (1976[b], 21).[1] Tying teleological descriptions of nonhuman behavior to dead metaphors affirming literal claims of consequence etiology provides an explanation for the fact that, although anything can be described in metaphorical teleological language, only some things, for example, the growth of plants but not the growth of crystals and the changing colors of a chameleon but not the changing colors of the sunset, can be described in literal teleological language.

The analysis of goal-directed behavior resulting from such claimed dead metaphorical extension is as follows:

S does B for the sake of G iff:
(i) B tends to bring about G.
(ii) B occurs because (i.e., is brought about by the fact that) it tends to bring about G. (Wright 1976[b], 39)

The term "tends" is interpreted very broadly, including "brings about, is the type of thing that brings about, tends to bring about, is required to bring about, or is in some other way appropriate for bringing about" (Wright 1976[b], 38-39).

Wright claims many virtues for his analysis. It accommodates the fact that teleological characterization is explanatory, for it talks about why the feature is present. It has the virtue of not being tied to any particular internal or external mechanism. It explains the forward orientation of teleological explanations, for "G" lies in the future. It reveals the difference between teleological explanations and what are sometimes called merely causal explanations. It provides a proper role for anthropomorphism in the discovery of teleological features, presumably because of its dead metaphor genealogy, while not allowing anthropomorphism to enter into the justification of teleological claims. Although it is not clear just how, Wright also claims that his analysis helps one see why the concept of trying is central to teleology and that it shows how teleological features can be said to be directly observed (1976[b], 55-56). One could hardly ask for more.

One is, of course, immediately troubled by the imprecision introduced into the analysis by the interpretation of "tends." The inclusion of the expression "appropriate" suggests either an element of intentionality, something firmly rejected by Wright, or, barring that, circularity, rejected by everyone. We think of something as appropriate when we project it against envisaged circumstances and it seems to us that it should help in reaching our goal, but that talk is in the intentional mode of plans, ideas, and projections.[2]

More important, however, is that the clearest examples of goal-directed behavior exhibit no consequence-etiology. The consequence of going to the store was buying the bread, not having the goal to buy the bread. Since the consequence, actually buying the bread, did not determine going to the store, it is incorrect to say that the consequence of going to the store determined going to the store, that is, it is incorrect to claim that the consequence of the action determined the action. Therefore, it is not accurate to describe the incident as an example of consequence-etiology. One could, of course, describe it as "intended-consequence-etiology," but that is something quite different. Since no one is likely to reject Wright's conventional analysis of the teleological statement "I went to the store in order to get some bread," doubt is, thereby, cast on his general thesis that teleology involves consequence-etiology. The thesis of consequence-etiology does not accommodate what is usually acknowledged as the paradigm case of teleological explanations, the explanation of intentional human action.

A resulting additional problem is that if reference to consequence-etiology is not present in teleological descriptions of human behavior, it can hardly be metaphorically transmitted to teleological descriptions of nonhuman behavior.

Furthermore, it was precisely the having of a goal or purpose in the example of a person buying bread that Wright utilized to remove the threat of reverse causation. If, when a teleological expression is metaphorically extended to nonhuman cases, reference to the having of a goal or purpose drops out, then the problem of reverse causation is reinstated, for the very element used to prevent the implication of reverse causation has been removed. Thus, although Wright expresses indignation with analyses that imply reverse causation, his own analysis is vulnerable to that charge. His analysis of teleological statements such as "The rabbit runs through the hole in the fence in order to escape the dog" removes the very element used to prevent the implication of reverse causation in such statements as "I went to the store to get some bread."

Another problem is that Wright, without supporting argument, simply declares that in the metaphorical extension to nonhuman subjects, the reference to purpose and goal drops out. Metaphorical statements that are also teleological include such statements as "The clouds chased each other across the sky" and "The wind searched out the cracks in the cabin wall." It is the suggestion of purpose and goal that gives these statements their point, for denying such reference produces incoherence. To say that the clouds chased each other but were not trying to catch each other or that the wind searched out the cracks but was not trying to find the cracks is self-contradictory. Expressions such as "trying to catch" and "trying to find" clearly refer to goals and the behavior directed to reaching them. Reference to purpose and goal in these metaphorical teleological statements did not drop out. There is no reason to assume without supporting argument that it would have dropped out in the metaphorical teleological statements that Wright believes were the ancestors of literal teleological descriptions of goal-directed behavior of nonhuman subjects.

A major problem with this formula is that it is not able to handle behavior directed to goals that are impossible to achieve. Although we have seen this problem in other analyses, because Wright's analysis is widely accepted, it seems especially important to examine this issue carefully.

Wright is aware that behavior can be goal-directed without the goal being reached. "Some of the clearest cases of goal-directedness consist in unsuccessful attempts and searches launched for a nonexistent goal" (1976[b], 29). The flexible linkage between the behavior and the goal attainment conveyed by the broadly defined "tends" allows for occasional goal-failure, but not for goals impossible to achieve. Historically prominent examples include Ponce de León searching for the fountain of youth and Somerset of Worcester trying to build a perpetual motion machine, as well as the less dramatic searches for caloric and ether. Ehring divides impossible goals into those that are physically impossible, such as behavior directed toward the goal of traveling faster than the speed of light, and those that are logically impossible, such as drawing an inference which does not logically follow (Ehring 1986[b], 127-31).

The class of goal-directed behavior involving impossible goals is neither small nor esoteric. Many, though not all, folk remedies belong. "Tie a bag containing the sufferer's nail paring to a live eel. It will carry the fever away." "Rub the wart with the skin of a chicken gizzard, then hide the skin under a rock. The wart will disappear." "To help hair grow, break a section of a grape vine, set in a bottle, and let the juice drain. Rub the juice in your hair" (Wigginton 1972, 238, 246, 248). Although these rituals have been performed in order to achieve those results, they are not covered by Wright's analysis because they do not reduce fever, remove warts, or grow hair, are not the type of thing that has those results, do not tend and are not required to do so, and are not appropriate for doing so.

Examples are not limited to unusual enterprises or even to human behavior. At whatever level directed behavior is accepted, examples directed to impossible goals lie close at hand. Pursued prey often make the wrong turn. If one regards rabbits as capable of goal-directed behavior, and Wright does, consider a pursued rabbit dashing left down a path that ends in a cul-de-sac instead of turning right on a path leading to dense undergrowth and freedom. Although it is legitimate to say, "The rabbit ran down the left path in order to escape," that behavior is not the type of behavior that brings about safety, does not tend to bring about safety, and is neither required nor appropriate for bringing about safety. Arthur Minton writes of an animal in a laboratory.

> We can imagine, for example, that it is the subject of an experiment and that the experimenters will continue to frustrate the animal's attempts to secure food until it expires. In this situation it is a fact that none of the animal's behavior will result in the achievement of the goal. Clearly the behavior will not tend to bring about the goal, and, hence, even the minimal condition (i) is not satisfied. Nevertheless, it is obvious that the animal is "trying to get" the food, is actively doing something to achieve the goal; and yet it cannot be goal-directed by Wright's analysis. (Minton 1975, 303)

Indeed, Wright's own examples do not work. Regarding fleeing rabbits, he says, "The fence might be too far away to reach safely, or the hole might be too small, or even covered by plate glass," believing that his broadly defined "tends" covers such cases (Wright 1976[b], 38). If, however, the fence is really too far away, the hole really too small, or the hole firmly and completely covered with stout glass, the behavior of running toward such a hole or toward a fence that is too far away does not even tend to realize the goal. Although running toward holes in fences tends to effect escape and is appropriate behavior, running toward too distant fences or too small holes or toward holes covered with stout glass is fatal. One must resist the temptation to say that the rabbit "thought" he could reach the fence or "expected" that he could get through the hole, for such language brings in intention and mental states and would require rewriting the analysis of teleology in a way Wright takes pains to avoid.

Curiously, in an earlier article, Wright criticized Braithwaite's plasticity theory along these very lines:

> It does not seem unreasonable to suggest that there might be occasions on which we would say a system manifested purposive behaviour even though, given its initial state, it was nomically *impossible* for it to achieve the goal to which it aspired. Behaviour might be purposive even though there are *no* possible gamma-goal-attaining causal chains. The fish struggling to get free of its net would seem to be a case of this type. It is struggling toward its goal of freedom even though achievement is impossible merely in virtue of its initial state and the relevant laws. (Wright 1968, 217)

In general, probability of success, even low probability, can always be eliminated from a constructed example by making the example such that the behavior convincingly, thoroughly, and consistently fails because success is either logically or physically impossible. With probability of success zero, it is incorrect to say that the behavior "brings about, is the type of thing that brings about, tends to bring about, is required to bring about" or is "appropriate" for bringing about the goal in question. Although Wright's analysis can handle ordinary goal-failure, contrary to the opinion of the many who accept Wright's analysis, it cannot handle extreme goal-failure, that is, behavior directed toward impossible goals.[3]

The central difficulty in understanding teleology is, of course, understanding how one can explain an item of behavior by referring to something that happens later or not at all. Wright does it by relating the behavior to what it produces, tends to produce, what requires it, etc. This situation, itself, has an atemporal status, and so applies also when the behavior occurs. However, in explaining his analysis, Wright insists, as he must if teleological language is to be given an empirical interpretation, that the explanation is purely causal. "Given the force of 'because' and 'brings about' (and 'etiology') in this analysis, the demonstration that a bit of behavior, B, occurred because it would tend to bring about some goal G is methodologically indistinguishable from the demonstration of standard, orthodox causal links" (Wright 1976[b], 41). Thus, a teleological explanation is not an alternative to a causal explanation but is rather a special kind of causal explanation. "A merely causal explanation of B would provide an etiology in terms of the *antecedents* of B, not its consequences. The causal teleological contrast is *among* etiologies, not between etiologies and something else" (Wright 1976[b], 57). "Merely causal" explanations would be such common ones as "The vase cracked because the water froze" and "The patient was irrational because of mercury poisoning." In contrast, the teleological variety of causal explanation has a forward orientation because it talks about what something tends to produce as the reason why it occurs, but the reason, Wright insists, is still a cause. "So B occurs *because* of what will ensue, but the statement of the cause is always appropriately put in the future tense: that things were such that G will (tend to) ensue" (Wright 1976[b], 56).

The difficulty with this is that, even if it were objectively established that the behavior in question produces a certain result, is required for it, or is appropriate for it, that, by itself, does not explain why the behavior occurred. Wright requires, as said above, that "because" in the second line of the analysis,

> (ii) B occurs because (i.e., is brought about by the fact that) it tends to bring about G

be interpreted causally. Therefore, it is more perspicuously rendered,

> The fact that B tends to cause G causes B.

Using Wright's earlier example, consider the teleological statement,

> I went to the store in order to get some bread,

which, following his formula, is analyzed as,

(i) Going to the store tends to bring about getting bread, and
(ii) Going to the store occurs because it tends to bring about getting
   bread.

Since "because" is to be understood causally, this is to be understood as,

(i) Going to the store tends to bring about getting bread, and
(ii) The fact that going to the store tends to bring about getting bread
   causes going to the store.

It is incorrect to claim that the mere fact that going to the store makes
it more likely that one gets bread causes one to go to the store. After all,
one may not know that the store sells bread or that there is a store there
at all, or one may know all that but not want any bread or be unable to
go. Indeed, there may be things in the store that no one would willingly
get, such as rotted fruit. It would be a strange understanding of
goal-directed behavior to say that the behavior in question in such cases
depends on not knowing the link between the behavior and the situation
it produces.

The situation is not limited to human behavior. Wright speaks of a
rabbit running in order to escape the dogs. His analysis, with the causal
interpretation of "because" made explicit, would be,

(i) Running from the dogs tends to bring about escape, and
(ii) The fact that running from the dogs tends to bring escape causes
   running from the dogs.

This seems plausible until one notes that flying also tends to bring
about escape, but it is not suggested that this fact causes rabbits to fly.
Clearly, much is being assumed that is not stated. In addition to the
requirement that the behavior be within the animal's capacities, the final
state must be of a certain kind, a kind which includes survival, pleasure,
and avoidance of pain, for we note that, contrary to the formula, the fact
that dashing into a lake tends to cause drowning does not cause the rabbit
to dash into a lake, and that is within its capacity. Further, if the rabbit
does not see the dogs coming, the fact that running from them tends to
bring about escape has no effect on its behavior.

The relation of behavior producing a certain effect, tending to produce
that effect, being required for that effect, or being appropriate for bringing
it about encompasses a range of conditions far too large to serve as
causes of behavior. Although chartering a plane to the South Pole tends

to bring about getting there, that fact does not cause that behavior to occur, for the fact is true for everyone, but the behavior is rare. Dumping heavy metals into the waterways tends to bring about poisoning the water supply; yet that fact, far from being a cause of the behavior in question, is a deterrent to it. Because examples can be endlessly generated having behavior that is impossible to perform or that no one would want to perform, the fact that the behavior is related to a goal in the various ways allowed by Wright is not a cause even in the modest sense of a contributing condition. This means that even with the statement asserting goal-directed behavior being true, the second line of the analysis may be false, indeed, usually will be. It appears more plausible in its original form, "B occurs because it tends to bring about G," than in the form "The fact that B tends to bring about G causes B to occur," because the former is read as a reason and the latter is read as a cause. The sharp difference suggests that reasons are not causes.

Although Wright does not analyze teleological behavior in terms of behavior patterns but rather in terms of causal links, he insists that the relations expressed in his analysis are empirically and publicly observable. It can, he claims, be known by experience that certain behavior, B, tends to produce a certain result, G, and also that B occurs because it tends to produce that result. He does not, however, mean that empirically determining these causal relations is infallible, but only that it is as good as our ability to determine causal relations generally: "The demonstration that a bit of behavior, B, occurred because it would tend to bring about some goal G is methodologically indistinguishable from the demonstration of standard, orthodox causal links" (Wright 1976[b], 41). To demonstrate that a certain B occurs because it brings about G requires the elimination of alternative accounts.

> If we suspect that there is something in the order in which the different escape routes succeed each other that influences the occurrence of the particular Bs, we may randomly change that order and repeat the sequence a number of times. To systematically eliminate other ranges of alternative accounts we could change the colors, substances, shape, and sizes of the environmental objects (e.g., those that define the escape route). (Wright 1976[b], 42)

Such elimination may produce either a single B or a class of Bs, any one of which meet the terms of the analysis.

Alicia Roque identifies a difficulty with this, however, in that determining whether the terms of the analysis apply may require a number of trials. Furthermore, many cases of goal-directed behavior do not allow for correcting procedures. One such kind is behavior done in order to achieve G when it does not tend to produce G but still produces G on this occasion. "A good example of this might be: *S*, a student with nothing to lose, flatters his teacher to get a higher grade even though *S* is well aware that flattering is for the most part the wrong sort of thing to try, and in particular with this teacher. Yet this time it works!" (Roque 1981, 157). If one performed tests on such behavior, the tests would show that such behavior does not tend to produce higher grades, but using such test results to interpret the goal of the above example would produce an error.

Furthermore, some actions are extremely uncharacteristic of the agent. Roque's example is a person of a kindly and peaceful temperament on one occasion firing a gun with intent to kill. Because the action is so out of character, the ordinary indicators are not taken at face value, and since it is a one-time occurrence, no pattern, even a new pattern, is there to be discovered.

Another counterexample is that of an agent changing his goal while performing the same behavior, as in performing a task at first to win the prize and then to please the examiner. Yet another is that of pretending behavior, performing B to appear directed to G when it is really directed to H. Roque's final example is that of an agent making his behavior appear nonteleological when it really is teleological, as one might do to thwart an experimenter (Roque 1981, 156-60). All these examples have in common the feature that determining the goal is blocked because additional testing is rendered useless.

> Contrary to Wright's claim, then, reference to behavioral dispositions alone explains nothing and reference to mental items themselves is necessary in order to distinguish the behaviors. A strictly empirical, behavioral account will not do because dispositions to behave a certain way are not uniquely correlated with mental items. (Roque 1981, 161)

Wright analyzes functions in a way parallel to his analysis of goal-directed behavior, that is, in terms of consequence etiology. Functions are consequences: "The function of something, X, is always some consequence of X's being there (wherever) or of its having a certain form" and "The function is a consequence of something with precisely that form being there" (Wright 1976[b], 77). In addition to a function's being a consequence of X, ascribing a function to X also explains why X is there. "The function of X is that particular consequence of its being where it is which explains why it is there" (Wright 1976[b], 78). This is clear, he notes, from the fact that "What is the function of X?" and "Why is X there?" are both answered by the same reply (Wright 1976[b], 80); "merely ascribing a function is to offer a consequence etiological account of the existence or form of the thing with the function" (Wright 1976[b], 91). Thus, the two central features of ascriptions of goal direction are also central features of ascriptions of functions: to ascribe is to explain and the explanation is by attributing consequence-etiology. "Function attributions are teleological explanations" (Wright 1976[b], 91). Wright does an outstanding job of arguing for the position that functional descriptions are explanations why the property is there and effectively criticizes the contrary views of Beckner, Hempel, and others.

Wright's formal analysis of functions, reflecting these two features, is as follows:

> The function of X is Z iff:
> (i) Z is a consequence (result) of X's being there,
> (ii) X is there because it does (results in) Z. (Wright 1976[b], 81)

The curious phrase "is there" is meant to be short for "has the form it has," "exists," etc. "As remarked above, 'is there' is a rather general place marker that takes on different significations in different sorts of cases, and in a way that is usually clear" (Wright 1976[b], 81). To illustrate, borrowing his example, "The function of quills on a porcupine is protection from predators" is analyzed as "Protection from predators is a consequence of the quills' being on the backs of porcupines, and quills are there because they protect from predators." In contrast, bilateral symmetry is not there because of what it does, and it also has no function (Wright 1976[b], 91-92).

Michael Levin accepts Wright's analysis and puts it to practical use in identifying the functions of artifacts in archeology. An already existing hole comes to be used for catching game. "But is the hole *for* catching game; is it a trap? The answer certainly seems to be 'no' if one contrasts this fortuitous hole with a physically similar hole that another group of hunters dug *in order to* catch game" (Levin 1976, 229). Merrilee Salmon, describing how archeologists might identify certain smooth river stones as stones used for polishing pottery, expresses well the flexibility Wright must have had in mind.

> Although these stones are selected for this purpose and are kept and passed along from mother to daughter for this use, they certainly do not

come into existence, even in the weak sense of being modified in some way, because someone figured out that they would be good for polishing. L. Wright's formulation of the definition of function is sufficiently broad to avoid this difficulty: the object *is there* because it does (results in) the function. His use of "is there" is deliberately vague to accommodate a number of different sorts of situations, such as this one in which natural objects are specifically selected, though not modified, to do a particular job. (Salmon 1981, 23)

An interesting feature of this formula is that, since it does not specify the consequence, Z, which explains why X is there, although Z may be survival or reproductive efficiency or linked to it, it need not be. This feature of the formula is intended by Wright to avoid a defect he saw in several analyses, especially in that of Ayala, that by limiting the consequence to survival or reproductive efficiency, it rules out design on mere linguistic grounds.

And this seems to suggest that it is impossible by the very nature of the concepts—logically impossible—that organismic structures and processes get their functions by conscious intervention (design) of a Divine Creator. This, I think, is an analytical arrogance. I am, personally, certain that the evolutionary account is the correct one. But I do not think this can be determined by conceptual analysis: it is not a matter of logic. (Wright 1976[b], 96-97)

A major problem for anyone who analyzes functions is how to give an account that will do justice to the functions of human artifacts while covering the functions of the parts and processes of organisms. Some, such as Braithwaite, Nagel, Canfield, Lehman, Ruse, and Neander, simply ignore the problem. Wright, along with Wimsatt and Woodfield, is to be credited with taking on the harder task of producing a comprehensive approach. Wright observes,

Functional ascriptions of either sort have a profoundly similar ring. Compare, "the function of that cover is to keep the distributor dry," with, "the function of the epiglottis is to keep food out of the windpipe." It is even more difficult to detect a difference in what is being requested: "What is the function of the human windpipe?" vs. "What is the function of a car's exhaust pipe?" (Wright 1976[b], 106)

Wright divides functions into conscious functions and natural functions. Conscious functions are those involving intention; natural functions are those not involving intention. The difference indicates a difference in origin, not in meaning, for one analysis serves both. Where the function comes from is "a matter of mere etiological detail" (Wright 1976[b], 97).

Wright holds, as does Wimsatt, that both conscious and natural functions have their origins in selection: "Functional explanation depends essentially on a selection background" (1976[b], 101). Two kinds of selection are involved, conscious selection and natural selection. "For just as conscious functions provide a consequence-etiology by virtue of conscious selection, natural functions provide the very same sort of etiology as a result of natural selection" (1976[b], 84).

Wright sees it as a common fault that analysts are so concerned to make the analysis fit natural functions that "these formulations are nearly impossible to apply to the conscious cases" (1976[b], 106). He, instead, begins his analysis with conscious functions and sees the concept of natural selection as an extension of the concept of conscious selection (1976[b], 85). The paradigm case of selection involves conscious choice, especially choice based on reasons, as in his example "They selected DuPont Nomex because of the superior protection it affords in a fire" (1976[b], 85). However, he sees the difference between conscious selection and natural selection as only "the slightest change in nuance" (1976[b], 84) and warns the reader that it would be "obscurantist . . . to drive much of a conceptual wedge between conscious and natural consequence-selection" (1976[b], 87).

The term "conscious," however, seems ill-advised. The taxonomy of conscious and natural functions leaves out artifacts produced by habitual action, such as much assembly line production. Anyone who has worked in a factory knows that, because of overlearning, endless repetition, and boredom, much of the shaping, fitting, sorting, painting, and packaging are performed, and performed properly, with attention directed elsewhere. Wright is aware that not all human selection is conscious, for he immediately makes provision for subconscious and what he calls "nonconscious" human selection. It surely, however, is misleading calling a kind of function "conscious functions," which includes a large subclass in which what is done is done unconsciously. Furthermore, Wright allows the extension of paradigm selection to form "a spectrum from more or less literal to openly metaphorical" (1976[b], 84-85), which certainly raises questions about how consciousness fits in with metaphorical selection.

Wright has been criticized for requiring that one knows the etiology of an item if one ascribes a function to it. Nagel notes that Harvey discovered that circulating the blood was a function of the heart even though he did not know the etiology of the heart (Nagel 1977, 284). Boorse makes a similar point.

> My first criticism of the etiological interpretation of biological function statements is that it is historically implausible. The modern theory of evolution is of recent vintage; talk of functions had been going on for a long time before it appeared. When Harvey, say, claimed that the function of the heart is to circulate the blood, he did not have natural selection in mind. Nor does this mean that pre-evolutionary physiologists must therefore have believed in a divine designer. (Boorse 1976, 74)

Such criticisms are misguided. Wright's etiological analysis says nothing about the details of the etiology; it is not, therefore, necessary to know such details to claim that the item having the function is there because of the function. It is not necessary to know whether the item having the function came about by natural selection, design, or in some other way to know that, however it came about, it is now present because it does so and so. The situation is similar to ordinary causal claims. To claim that an illness has a cause is different from claiming what its cause is, and one can legitimately assert the first while remaining ignorant about the second.

Wright holds that natural selection is selection without a selector. "We might want to say that *natural* selection is really *self*-selection, that nothing is *doing* the selecting; given the nature of X, Z, and the environment, X will *automatically* be selected" and "And as we have seen, selection does not require a selector" (Wright 1976[b], 86, 105). However, to select is to choose, as Wright, himself, acknowledges, and involves discriminating and picking out. Such actions require agents, and this not merely for grammatical reasons. Voting, for example, is a kind of choosing, expressed in a certain manner and occurring in a certain context. Voting cannot occur unless there is a voter. Ballots allegedly cast by certain individuals, who, upon investigation, turn out to be deceased or were incorrectly listed as residents of vacant lots, are not counted. Given X (such as green coloration) and Z (such as decreased visibility) and the environment (such as one that includes green foliage and predators), saying that selection occurs automatically says merely that given X and Z and the environment, the type of organism in question that has property X tends to survive. Although it is true that there is no selector, there is also no selection.

The fact that selection requires a selector and that natural selection has no selector seems sufficient evidence to conclude that the term "selection" in the expression "natural selection" is used metaphorically. As noted earlier, Darwin, himself, used the term that way, pointing out that what happens in nature when organisms with certain properties have different survival tendencies can be compared to what happens when agricultural breeders select offspring with certain characteristics for reproduction and reject others. What is similar is not the process but the result. The result in both cases is that the number of organisms with a certain characteristic increase and those without decrease.

As noted earlier, to select is to choose, to elect, to pick out. Although choosing does not always require reasons, for there is such a thing as arbitrary choice, the ability to choose does seem to presuppose the ability to have reasons, and having reasons has traditionally been taken as a distinguishing mark of mind. Saying that nature selects which organisms survive is surely no different from saying that the wind selects which trees to uproot and is unmistakably metaphorical. Darwin, himself, at least sometimes, acknowledges its metaphorical status: "In the literal sense of the word, no doubt, natural selection is a false term; but who ever objected to chemists speaking of elective affinities of the various elements?—and yet an acid cannot strictly be said to elect the base with which it in preference combines." He justifies the use of such terms by acknowledging their metaphorical status. "Every one knows what is meant and is implied by such metaphorical expressions" (Darwin 1878, 63). The expression "natural selection" is closely related to the expression "struggle for existence." Indeed, both occur in the full title of *The Origin of Species*. Since these expressions are applied also to plants and since one of them, "struggle for existence," surely must be applied to plants metaphorically, the other, "natural selection," must be as well. Darwin is admittedly not consistent, however, for he does not distinguish the struggle for existence of an animal attacked by a predator from the struggle for existence of seeds carried by the wind.

Wright acknowledges the metaphorical origin of "natural selection" in his consequence-etiological analysis of goal-directed behavior. "But it is the consequence-etiological form that the paradigm and the metaphorically extended cases have in common" (1976[b], 73). He also acknowledges the metaphorical origin of "natural selection" in respect to natural functions: "The logical structure of natural functions is best understood by noticing its intricate parallels with conscious functions: by seeing that structure as issuing from the death of an explanatorily rich metaphor" (1976[b], 106). It is interesting, however, that reference to metaphorical origin gets scant mention in his analysis of functions in contrast to the importance it assumed in his analysis of goal-directed behavior. Since any two things, no matter how different, can be related metaphorically, if "selection" in "natural selection" has only a metaphorical relation to human selection, there is little to support Wright's claim of having provided a single analysis of functions in which the shift in meaning in going from conscious functions to natural functions involves only the "slightest change of nuance."

Wrights's view that natural selection is a case of genuine selection and is very close to conscious selection is encouraged by construing natural selection as creative and not simply a force that destroys. To this end, Wright approvingly quotes Ayala, who rejects the idea that natural selection acts merely in a negative way as a kind of sieve and attributes creative power to it (Wright 1976[b], 86-87).

As noted earlier, Wright's statement that natural selection says that "given the nature of X, Z, and the environment, X will *automatically* be selected" means simply that given the nature of X, Z (where Z is a result of X), and the environment, things having X will tend to survive. This interpretation accepts a large range of events that no one wishes to describe in terms of natural selection. Given the nature of granite and its consequent resistance to abrasion and given an environment of turbulent water, boulders made of granite tend to survive, yet no one says that some boulders are made of granite because of natural selection. In spite of warm weather, given the nature of the property of lying in deep valleys and the consequent property of being shielded from the sun and given an environment of warm weather, snow tends to survive. Yet we do not say that the property of lying in deep valleys is the result of natural selection. Clearly, more is involved in natural selection than is conveyed by Wright's brief account, such as X's referring a heritable property in a living organism.

Even aside from such corrections, Wright seems wrong in attributing creativity to natural selection. The more fleet survive by natural selection acting on a mixture of the more and less fleet and destroying the less fleet. Indeed, agricultural breeding, which is the source of the idea of natural selection and even involves conscious selection, is, itself, purely destructive. Whatever creativity is present must come from other sources, such as mutation, and that is not a component of natural selection. Cummins expresses it well: "We could, therefore, think of natural selection as reacting on the *set* of plans generated by mutation by weeding out the bad plans: natural selection cannot alter a plan, but it can trim the set" (1975, 751). Nagel makes the same point: "*Which* of its genes an organism transmits to its progeny, is determined by random processes that take place during meiosis and fertilization of the sex cells. It is *not* determined by the *effects* that the genes produce in either the parent or daughter organisms" (1977, 286). It does not seem far from the mark, therefore, to characterize natural selection as a negative process, acting somewhat as a sieve, and not creative.

Cummins points out that since natural selection does not determine what features occur but only which of the occurring features survive, natural selection does not explain the presence of organs or features having functions. This undercuts Wright's consequence-etiological analysis of functions for natural systems, since that analysis appeals to natural selection etiology. Cummins believes that function statements do not explain the presence of the item having the function except in the case of artifacts.

> For it seems to me that the question, "why is *x* there?" can be answered by specifying *x*'s function only if *x* is or is part of an artifact. . . . the viability of the sort of explanation in question should depend on the assumption that the thing functionally characterized is there as the result of deliberate action. If that assumption is evidently false, specifying the thing's function will not answer the question. (Cummins 1975, 746-47)

Kristin Shrader-Frechette argues that Wright is in error when he and others "assume that natural selection operates to produce organs of a particular kind *because* their presence gives rise to certain effects. . . . Since mutation of genes as well as which genes are transmitted to offspring are determined by random processes, natural selection does not operate so as to give rise to certain effects" (1986, 85).

Wright's goal-directed behavior analysis reads in part "B tends to bring about G," while his function analysis reads "Z is a consequence (result) of X's being there." The fact that his function analysis does not read "Z tends to be" seems to indicate that Z, to be a function of X, must actually be a consequent or result of X. There are two kinds of cases in which the consequences do not occur, one trivial, one not. In the trivial case, Z does not occur simply because X does not occur or is not used.

> Thus it is worth noting that in some contexts X can be said to do Z even though Z never occurs. For example, the button on the dashboard activates the windshield washer system (that's what it does, I can tell by the circuit diagram) even though it never has and never will. All that

seems required is that X be *able* to do Z under the appropriate
conditions: in this case, when the button is pushed. (Wright 1976[b], 81)

The point seems to be that functions are dispositional properties.

The other case is not trivial. There are many examples in which Z
cannot occur, that is, does not occur even when X occurs. Achinstein
notes, "There is another type of example, however, which appears to
impugn both of Wright's conditions as necessary conditions. Artifacts can
be designed and used to serve certain functions which they are incapable
of performing" (1977, 349). Indeed, Wright, himself, is aware of the
problem that functions are not limited to consequences that are possible.
"If the windshield washer switch comes defective from the factory and is
never repaired, we would still say that its *function* (italics) is to operate
the washer system—which is to say: that is what it is *supposed* to do"
(Wright 1976[b], 112).

The problem of how to analyze functions in which Z cannot occur is
not limited to conscious functions. "Illustrations among natural functions
are as easy to find. Pointing to a defective epiglottis, I might say, 'the
function of that flap is to keep food out of the windpipe, although this one
can do nothing of the kind.' This also may be paraphrased: that is what
it is *supposed* to do" (Wright 1976[b], 112). Wright claims that we flag
such function talk in such cases by emphasizing, italicizing, or
underlining to indicate variant status. Not so. Function talk about
defective artifacts and organs is not normally marked in any way, which
raises doubts about his claim that they are variants. Function ascription
is often in normative language, for example, "The switch is supposed to
turn on the light," and that is said whether or not it turns on the light. We
freely ascribe functions to artifacts and organs not even knowing whether
they work. We could not do this if Wright's claim about flagging and
deviant meanings were correct.

In response to examples of natural functions in which X does not result
in Z, Wright again argues that the analysis is about types, not instances.
On this view, even though this particular epiglottis does not and cannot
keep food out of the windpipe, its function, nevertheless, is to do just that
because the epiglottis typically does this. "For the natural cases, the
variation is accommodated by noticing that when talking of an individual
case 'X does Z' should be read 'Xs do Z'; and hence the change required
is simple: 'Xs *typically* do Z'" (Wright 1976[b], 113).

One could, perhaps, use the type-argument also for some conscious
functions, namely, those involving an occasional defective artifact in a
normal run of proper ones, as might occur in assembly line production,

but Wright does not do so. In any event, that maneuver would hardly avail for artifacts so poorly designed or constructed that they could not work. Cure-all patent medicines would be an example of the first and trashy manufactured goods of the second. In addition, there are individually constructed artifacts concerning which there are no defective types to discuss. Thus, one may make an individual birdhouse for purple martins and, by mistake, construct it with the hole too small, making it impossible for such birds to enter. The function of X, the hole, is Z, to enable purple martins to enter and exit, although it is nomically impossible for X to result in Z. Examples of this kind involve the same problem discussed earlier concerning behavior directed to impossible goals.

Wright responds by saying that in such cases, the formula, "Z is a consequence (result) of X's being there" or simply "X does Z" means that some person supposed that X does Z.

> In order to provide a conscious consequence-etiology in terms of Xs resulting in Z, when X does not result in Z, it is necessary to resort to the other sense of "suppose": *somebody* supposed that X would result in Z, *that's* why it is there. So once again, the etiology is virtually the same as if the thing had worked, it just did not work. (Wright 1976[b], 113)

Wright claims that "suppose" had two senses, the one referred to in the previous quotation and "that famous sense in which 'X is supposed to happen' does not imply that anybody actually supposes X *is* going to happen" (1976[b] 112). Something has gone awry. "To suppose" means about the same as "to expect" or "to assume." Suppositions, expectations, and assumptions are not part of impersonal settings. Expectations require subjects who have them, and suppositions do likewise. Wright is misreading idiomatic expressions, such as "December is supposed to be colder than November" or "Exercise is supposed to improve health." No supposer is mentioned because the speaker believed most people suppose these things. After all, one could have said, "Exercise is thought to improve health" and no one would have concluded that there are thoughts without thinkers. In a universe without beings of a certain level of mental life, there would be no expectations, assumptions, or suppositions.

In saying that conscious functions in which X does not produce Z are to be understood as saying that someone supposes that X would produce Z, Wright has brought intentionality back into the analysis. What he says is no doubt true and appears, moreover, to be the correct way to understand such functions, but there is no way such an analysis can reasonably be derived from his formula. His formula says,

(i) Z is a consequence (result) of X's being there,
(ii) X is there because it does (results in) Z,

while his explanation of conscious functions in which Z is not there even when X is, that is, in which Z is not a consequence of X's being there, is, as we saw above,

somebody supposed that X would result in Z,
that's why it is there.

Briefly, his formula says that X causes Z while the above revised analysis says that Y thought X would cause Z. One is a causal claim; the other is not. The difference between the two is nothing less than that which separates the thought of something from that something. Saying that claiming the function of X is Z because someone supposed X would result in Z effectively removes any consequence-etiology from Wright's account, a point parallel to one made earlier about his analysis of goal-directed behavior. To bravely say that "the etiology is virtually the same as if the thing had worked" does not alter the situation. If supposition about a reality claim can be considered merely a variant reading of the reality claim, discourse has been undercut and communication falters. The repeated assertion that flagging occurs in such cases, "so the logic includes an emphatic flag for use at this borderline," is not accurate (Wright 1976[b], 113).[4] There is no such general practice.

Unlike several writers, Wright sharply separates functions from goal-directed behavior and argues that functions have no goals. "Clearly function and goal-directedness are not congruent concepts. There is an important sense in which they are wholly distinct" (1976[b], 74-75). Being goal-directed, he says, is a property only of behavior while being a function can be a property either of behavior or of something else. "'Goal-directed' is a behavioral predicate. The *direction* is the direction of behavior. . . . Conversely, many things have *functions* (e.g., chairs and windpipes) which do not behave *at all*, much less goal-directedly" (Wright 1976[b], 74).

This establishes only that goal-directedness and functions are not identical. To further establish their separateness, he points to behavior that has a function but no goal direction, "e.g., pacing the floor or screaming out in pain" and to behavior that has both a function and a separate goal.

"For example, some fresh-water plankton diurnally vary their distance below the surface. The goal of this behavior is to keep light intensity in their environment relatively constant. . . . But the *function* of this behavior is to keep the oxygen supply constant, which normally varies with sunlight intensity" (Wright 1976[b], 74).

Wright claims that experimenting with artificial light would enable one to determine whether keeping light level constant is a goal or a function. Such experiments, however, would show only a pattern of covariation between light intensity and depth of swimming, which would be compatible with either a goal or a function claim. After all, it would be surprising if his formula fit one but not the other, since they both, as he reminds us, assert consequence-etiology.

Indeed, the differences between his analysis of goal-directed behavior and his analysis of functions seem evanescent. One formula links its two elements with "tends" and the other links its two with "does" but adds commentary that relaxes that to "tends." In one, the first line has the order of "C does E," while, in the other, the first line has the order of "E is done by C." One uses the expression "is there," but the commentary interprets that so broadly that anything grammatically acceptable can be substituted. Finally, one uses the letters "B" and "G," while the other uses the letters "X" and "Z." However, since "X" may stand for behavior as well as "B," and "G" and "Z" both stand for consequences of the other element, there is no way, when one is talking about behavior, to determine which are the appropriate letters until after it is determined whether the example is one of goal direction or of function. One could interpret the behavior of the plankton in a way opposite to Wright's way and do so consistently within his analyses. Thus, whereas he says that keeping light intensity constant is a goal, the example fits his function formula as well:

> The function of diurnally varying the depth (X) is maintaining constant light intensity (Z) iff:
> (i) Maintaining constant light intensity (Z) is a consequence (result) of diurnally varying the depth (X),
> (ii) Diurnally varying the depth (X) occurs (is there) because it does maintain (results in) constant light intensity (Z).

Similarly, while he says that keeping oxygen supply constant is a function and not a goal, the example fits his goal-directed behavior analysis as well:

Some freshwater plankton (S) diurnally vary their depth (B) for the sake
of maintaining oxygen supply constant (G) iff:

(i) Diurnally varying the depth (B) tends to maintain the oxygen supply
constant (G).

(ii) Diurnally varying the depth (B) occurs because it tends to maintain
the oxygen supply constant (G).

Woodfield, in a review of Wright's book, notes that his two analyses
are not really different. "Ultimately, however, Mr. Wright's synthesis of
the two analyses is doomed, because it obliterates the initial distinction
between goals and functions. The (T) and (F) formulae do not differ in
any respect that entails a difference in the logic of the two kinds of
explanation" (1977).

Since Wright expects that the reader will agree on what is the goal and
what is the function and is probably correct in that expectation, and since
his formulae do not differentiate goals from functions when one is talking
about behavior, one wonders upon what such presumed agreement in
discrimination is based. The situation certainly suggests the traditional
way of differentiating between goal-directed behavior and function
behavior by appealing to internal direction and external direction. On this
view, one presumes that plankton must have some low-level awareness
and can probably distinguish levels of light and that they have some
preference for a certain level, making the source of the direction internal
to the subject behaving. On the other hand, we judge that plankton
probably cannot detect levels of oxygen. However, because we know they
would die without adequate oxygen and since the required level is
somehow maintained, we hypothesize that the direction is outside the
subject behaving. This view, of course, treats teleological statements
anthropomorphically, but historically, that is how they were treated and
is, of course, why they were controversial. Consequence-etiology does not
differentiate goal-directed behavior from function since it is found equally
in both.

The position that functions have no connection with goals is especially
difficult to maintain for conscious functions. Consider a roofer who puts
a roof on a house in order to keep out the rain. The action is goal-
directed and keeping out the rain is the goal. However, the roof also has
a function, and that function is the same as the goal of the behavior,
namely, to keep out the rain. The description of the goal-directed behavior
and the description of the function are about the same roofer, the same
roof, and the same rain. It seems unreasonable that these two descriptions
could have a different ontology. Indeed, one could rephrase both

descriptions with no change in meaning, using the single term "purpose" to refer both to the goal and to the function.

> The carpenter puts a roof on the house in order to keep out the rain

can be rephrased as

> The carpenter puts a roof on the house for the purpose of keeping out the rain,

and

> The function of the roof is to keep out the rain

can be rephrased as

> The roof is on the house for the purpose of keeping out the rain.

Since Wright considers the first sentence as being about behavior directed to a goal and since the second sentence has the same meaning as the first and uses the term "purpose," a term usually understood as being roughly synonymous with "goal," and since the corresponding function statement is also equivalent to a formulation using the term "purpose," it does not seem unreasonable to interpret functions as, at least, in some way involving goals.

In his account of human goal-directed behavior, Wright speaks of a person having goals, and in his account of conscious functions of defective artifacts he speaks of a person having suppositions. Talk of suppositions certainly utilizes intentionalistic language and talk of goals seems to, as well. It, therefore, should not be rejected out of hand that descriptions of goal-directed behavior and descriptions of functions are both intentionalistic and that the difference dividing them is merely whether the intentionality is internal or external. After all, Wright does acknowledge that there is an "enormous parallel that obtains between goals and functions" 1976[b], 80). Although he explains that parallelism in terms of his consequence-etiology, it could also be accounted for in terms of intended consequence-etiology.

As we saw earlier, regarding goal-directed behavior, the problem of reverse causation is avoided in human behavior, such as going to the store to get bread, by bringing in intentions, but when intention drops out in the metaphorical extension to the behavior of lower organisms, the problem of reverse causation is reinstantiated. With functions, the problem of reverse causation shows up again but receives different treatment. Most analysts, including Wright, agree that the clearest class of things having functions is that of human artifacts, such as hammers, saws, and umbrellas. In such cases, however, purposes and intentions play a prominent role. The reverse causation problem is easily resolved if reference to intentional states is admitted, for the function is then determined by intended consequences rather than actual consequences.

The analysis of functions says that X results in Z and Z is responsible for X's being there, does not refer to intentions and purposes, and, in the case of natural functions, positively excludes them. "The point of this chapter is to say something helpful about function attributions and functional explanations in instances not underwritten by human design or intent" (Wright 1976[b], 84). The first part of the analysis, "Z is a consequence (result) of X's being there," is openly about causal consequences or causal results; hence, it could be expressed simply as "X causes Z." The second part, "X is there because it does (results in) Z," is likewise a causal statement and could be rephrased as "X is there because it causes Z." Wright leaves "because" unanalyzed so that it can cover both conscious and natural functions. In the case of natural functions, since intentional properties are explicitly excluded, "because" must be interpreted causally. Thus, the second part of the formula, "X is there because it causes Z," when applied to natural functions, is to be understood as "X's causing Z causes X to be there."

One way to avoid implying reverse causation is to interpret the formula as "Z *was* a result of X's being there and X *is* there because X resulted in Z." An example would be "Protection from predators was a result of the quills' being there and the quills are there because they resulted in protection." Wright rejects this interpretation because it does not exclude vestigial organs:

> Use of the past tense in this way blurs the distinction between functional and vestigial organs, which is worth some pains to avoid in this context. Both kidneys and appendixes are there because of the function they had in the past; only kidneys are there because they do what they do, which is to say only kidneys (still) have a function. (Wright 1976[b], 89)

The solution to the problem of reverse causation, as Wright sees it, lies in insisting that function statements are about types, not about individuals, or, more accurately, that Z is always a type and that X usually is.

> The fly just consumed by Horatio the spider was not in any way
> involved in the etiology of Horatio's web-spinning ability. . . . when we
> say "the spider (or, a spider) possesses the ability to spin a web because
> that allows it to catch food," "the" and "a" are seldom used to refer to
> a specific individual, and "that" *never* does. "The spider" is usually
> equivalent to "spiders" (like, "the American farmer") and "that"
> invariably refers to a property (e.g., an ability or propensity) of a certain
> *type* of thing, and logically cannot be limited to a specific instance of
> the type. To make this explicit, the sentence above might be rendered,
> "spiders possess the ability to spin webs because web-spinning helps
> catch food." (Wright 1976[b], 88)

On the contrary, saying that ascribing functional properties to
individuals is seldomly done is to acknowledge that it is, nevertheless,
sometimes done. An example might be saying that the function of
Horatio's web is to catch insects. Further, to allow function ascription to
pluralities, as spiders, is to allow it to individuals, for pluralities are made
up of individuals and are not types.

In moving from the particular to the general, the properties and
relations among the particulars are retained. In moving from the particular
"Socrates is rational" to the general "Mankind is rational," the property
of rationality is retained, and, in moving from "This oak tree is taller than
that rose bush" to "Oak trees are taller than rose bushes," the relation of
being taller is retained. The properties and relations ascribed to types are
derivative, coming from the properties and relations ascribed to instances.
This is also the case when the relations are temporal or causal. The
temporal ordering of "Parents are older than their children" is derived
from the temporal ordering of statements describing individual parents
and their children, such as "Socrates was older than his children." The
causal ordering of "Lightning causes thunder" is derived from the causal
ordering in such statements as "That lightning stroke caused this peal of
thunder." Kind or type statements make general what is said in describing
instances; they do not create new relations or remove old ones. Thus, as
noted in the discussion of Neander (chapter 3), the causal relations that
obtain between X and Z as types must reflect those same causal relations
that obtain between the instances X_1 and Z_1, X_2 and Z_2, etc. Wright's
formula "X results in Z and X is there because it results in Z" must have
its counterpart in the language of instances.

Presumably, in the language of instances, the same Xs and Zs are
referred to, making the counterpart "X_1 results in Z_1 and X_1 is there
because it results in Z_1, X_2 results in Z_2 and X_2 is there because it results

in Z_2, etc." Since the term "because" cannot here refer to the having of a reason inasmuch as intentional states are excluded, it must refer to ordinary causation. Therefore, the above may be rewritten as "X_1 causes Z_1 and X_1's causing Z_1 causes X_1 to be there, X_2 causes Z_2 and X_2's causing Z_2 causes X_2 to be there, etc." However, because X_1's causing Z_1 occurs after X_1 is there and X_2's causing Z_2 occurs after X_2 is there, etc., this formula implies reverse causation.

Reverse causation could be avoided by rephrasing the above "X_1 caused Z_1 and X_2 is there because X_1 caused Z_1, X_2 causes Z_2 and X_3 will be there because X_2 causes Z_2, etc." However, the generalization of that would be "X resulted in Z and X continues to be there because it resulted in Z in the past," which is quite different from Wright's formula and reintroduces the problem of vestigial organs.

Thus, although Wright rejects reverse causation, saying, "But there is nothing in any of the ordinary ascriptions of goals or functions or motives or purposes or aims or drives or needs or intentions which requires us to reverse the normal cause-before-effect sequence" (1976[b], 10), his analysis does not escape that troublesome implication. Ruth Millikan, who prefers a historical approach, is one of the few who have noticed this rather devastating problem in Wright's analysis of functions: "Rather, we are asked to accept that *X* might be there *now* because it is true that *now* *X* does or *X*'s do result in Z. How the truth of [a] proposition about the present can 'cause' something else to be the case *at present* is not explained" (Millikan 1989[c], 299). Falk is acutely aware of Wright's difficulty, summarizing, "Obviously X's being there precedes X's causing Z. Either we have a contradiction or a claim of reverse causality. . . . Given the resources of his analysis, his generalizations may only be true in virtue of their instances" (1995, 338-39). Falk sees this as indicating the need for going to an analysis based on natural signs.

Furthermore, whatever the case may be in respect to natural functions, it is clear that statements about conscious functions may be about particular instances. Suppose one's car is in danger of slipping off the jack. A rock is jammed under a wheel. It is quite proper to say that the function of this particular rock under this wheel is to stop this car from rolling. Wright does seem to recognize that statements of conscious functions can be about individuals. Speaking of a banking lever on a furnace, he says, "Conscious selection can take place with respect to a specific instance and does not require generations as does natural selection" (1976[b], 88-89). As noted earlier, such functions are saved from reverse causation only by abandoning the consequence-etiological claim and talking about supposed or intended consequences instead of actual consequences.

Several commentators have leveled criticisms of Wright's analysis of function that fail. Patrick Grim imagines a situation in which an engine runs roughly because of a large gap in a piston ring. The vibration jars loose a pollution control valve that is drawn into the defective cylinder and lodges in the ring gap, making the engine run smoothly. Ignoring the unbelievability of the example, the valve "is there (at least in part) because it makes the engine run smoothly. . . . [and] the engine's running smoothly is a consequence of its being there" (Grim 1974, 63). Yet, making the engine run smoothly is not the function of the broken part. On the contrary, it is surely not obvious that the valve is in the ring gap because it makes the engine run smoothly. The only way this example can be rendered plausible is to interpret Grim's description as meaning that the valve remains in the ring gap because it made the engine run smoothly, separating the valve's initially breaking loose and settling there from its remaining there after settling there. When this is done, the example no longer fits Wright's schema, for in "Z is a consequence of X's being there," the phrase "X's being there" corresponds to the valve initially lodging in the gap, while in "X is there because it does Z," the phrase "X is there" corresponds to the valve remaining in after it has lodged there.

Boorse's challenge to Wright is similar: "Obesity in a man of meager motivation can prevent him from exercising. Although failure to exercise is a result of the obesity, and the obesity continues because of this result, it is unlikely that prevention of exercise is its function" (1976, 75-76). The part corresponding to Wright's "Z is a consequence of X's being there" is "Failure to exercise is a consequence of obesity," and the part corresponding to "X is there because it does Z" is "Obesity continues because of the failure to exercise." These two statements can be both true only if time sequence is read in, which makes the two references to obesity references to obesity at different times. Indeed, this is obvious even from the language, for one statement talks of obesity while the other talks of obesity continuing. The example does not fit Wright's schema, for in that schema, Z refers to the same item in both occurrences.

Another of Boorse's examples concerns a farmer confronted with a hornet in a shed. "The farmer's fright is a result of the hornet's presence, and the hornet's presence continues because it has this result" (1976, 75). Here again the language clearly reveals that two distinct conditions are substituted for Z—in the first Z, the farmer's initial fright, while in the second, its continuance—and so this example also does not fit Wright's schema.

Achinstein, in commenting on Wright's analysis of function, imagines a manager of a baseball team who keeps a player on the first team if and only if that player continues to bat over .300. "Jones is there (i.e., on the first team) because he continues to bat over .300. Jones' continuing to bat over .300 is a consequence (or result) of his being there (i.e., on the first team, which gives him practice and confidence)" (Achinstein 1975[b], 752). The batting over .300 that was responsible for Jones' being on the first team is not the same batting over .300 that results from his being on that team. Two different extended events are being referred to and so the example does not fit Wright's formula. Furthermore, since Achinstein's example is about conscious goal-directed behavior, it should have been applied to Wright's analysis of goal-directed behavior rather than of functions.

Ruse says that Wright's analysis of functions "seems to have a hole in it big enough through which to drive a philosophical coach and four," and goes on to explain: "The rain is there because the rivers run down to the sea (if the rain just vanished into the center of the earth we should eventually run out of water). The rivers run as a consequence of the rain's being there. But a function of the rain is hardly the river's running" (1979, 788).[5] Once again, time sequence is implied. For the statements to be true, either there are two separate runnings of rivers with rain between or there are two separate rains with a river running event between. This example cannot be substituted in Wright's X and Z, which refers to two things, not three.

Mark Bedau offers an example to counter Wright's analysis along similar lines.

> Consider a stick floating down a stream which brushes against a rock and comes to be pinned there by the backwash it creates. . . . part of the explanation of why it creates the backwash is that the stick is pinned in a certain way on the rock by the water. . . . the stick meets the etiological conditions: creating the backwash tends to pin the stick on the rock and the stick creates the backwash because doing so contributes to pinning it. (Bedau 1991, 648)[6]

If this description is taken as true, the reader must insert time sequence. The backwash that keeps the stick pinned for one moment is not the same backwash caused by the stick at that same moment. Swirling water takes time to swirl.

These several unsuccessful counterexamples are all based on ignoring the crucial time sequences hidden in their accounts. Wright's analysis, using a single variable, X, for both that which has the consequence and that which is there because of it, allows for none.

Peter Godfrey-Smith offers another counterexample. "Consider a small rock holding up a larger rock in a fast-moving stream. If the small rock did not support the larger rock, it would be washed away. Holding up the big rock is the thing the small rock does, that explains why it is there. So on Wright's original analysis this is the function of the small rock" (1993, 198). If the example explains anything, it is not why the small rock is there but only why it remains there once it is already there, and Wright's analysis is not about mere continuance. Further, the small rock remains there because of the downward force of the upper rock, not the upward force of the lower rock.

Wright's defense of teleological language in science is aggressive, comprehensive, and well-informed, and many now regard it as correct. As the examples above show, it stands up under criticism better than some have alleged. There are, however, a number of difficulties discernible, serious difficulties, which minor adjustments will not remedy.

Notes

1. Ruse ignores Wright's talk about dead metaphor and incorrectly claims that Wright's position is that teleological talk in nonhuman contexts is metaphorical. He says that Wright holds "that teleological explanation is metaphorical, being a transference from the human conscious case. We think in terms of ends in the case of human actions and intentions, and then we translate this kind of thought metaphorically to other situations" (1978, 199-200).

2. Although the reading "tends" is very broad and seems to be offered as a rough approximation, it is not so broad as some take it. Porpora believes it includes the behavior's having had a history of bringing about G. "We may say that 'B tends to bring about G' if B has a history of bringing about G, or, alternatively, if B, at present, is causally connected to G in such a way that B makes G more probable" (1980, 573). Ringen (1985, 571) concurs: "Wright's analysis does, of course, allow that goal-directed behavior can be explained either by laws of the form $B = f(H)$ or by laws of the form $B = f(T)$." By the former is meant laws that make the behavior a function of the history of the organism, such as past reinforcement. Though Porpora rejects the historical interpretation as in

error and Ringen accepts it as correct and needed, neither gives any reasons for claiming Wright intended it. The historical element in Wright's analysis, which appears most prominently in his account of functions, concerns past natural selection and has nothing to do with the history involved in respondent or operant conditioning.

3. Ringen (1985, 568) criticizes this position: "Wright's analysis does not require that goal-directed behavior make the future occurrence of the goal-state more likely." However, his position is based, with no textual evidence offered, on the unlikely reading of "B tends to bring about G" as including "B has a history of bringing about G."

4. Nagel (1977, 284-285) considers the statement that the function of the main spring in a watch is to power the rotating cogwheels. "On Wright's analysis, this is equivalent to saying that the spring does have this effect, and also that the spring is where it is *because* the spring has that effect. However, the second clause of the allegedly equivalent statement is surely an error. The spring was placed where it is by the manufacturer *not* because the spring is able to rotate wheels (as is required by Wright's analysis), but because the manufacturer *knew* or *believed* that this was so."

5. This example is also discussed in Ruse 1978, 201.

6. This example is also discussed in Bedau 1992[b], 786.

Chapter 6

Woodfield

Andrew Woodfield's book *Teleology* is a major contribution to the teleological debate and has contributed substantially to the understanding of teleological language. It is a richly detailed, perceptive, and informed work on a very complex and perplexing subject. For some reason, perhaps because it is not as reductive as desired, it has not been widely discussed, a fate it does not deserve. Woodfield begins with a historical sketch that provides perspective for what follows, evaluates several analyses of teleological language, concentrating on the plasticity theories of Braithwaite, Sommerhoff, and Nagel, offers valuable insights regarding cybernetic analyses, and finally provides a comprehensive and plausible analysis of his own.

In developing his own analysis, Woodfield regards as hopeless any attempt to construct a revised behavioristic theory, including one supplemented with reference to benefit. He reads the evidence as pointing toward an internalist view of goals, that goals are internal states of the agent rather than properties of the behavior, but toward an internalist position other than that of feedback. He notes that goal talk is usually accompanied by talk about what the subject perceives, believes, recognizes, learns, etc., implying that the subject has awareness (Woodfield 1976, 164-65). Borrowing from Ducasse, he suggests that goals are mentalistic, intentional objects, specifically, objects of desire. "There is no such thing as a goal to which the agent's behaviour is directed, yet which the agent does not have; and to have G as a goal is to have G as the object or content of a desire" (Woodfield 1976, 172). Again, "A goal just *is* the intentional object of the relevant kind of conception" (1976, 205).

Woodfield distinguishes having a goal from achieving a goal. Having a goal is a state of the subject; achieving a goal is a state involving both the subject and other things, a state involving the subject and the world.

It is only the former that is relevant to the analysis of goal-directed behavior. Whether or not Woodfield is correct in claiming that "goal" has only the sense he recognizes, there is no doubt that confusing "goal" as a state of an agent with "goal" as an object or feature of the world has been and continues to be responsible for much of the confusion in the teleology debate.

In developing his case for interpreting goals as intentional items, he notes that plasticity theorists, themselves, describe behavior in ways that imply mental states. Suppose a rat swims across a river and is rewarded with food. The behavior is described by saying the rat swam across the river in order to get the food. Woodfield now observes,

> The plasticity conditional "If the food were further to the left, the rat would swim further to the left" relies covertly on the assumption that if the food were further to the left, the rat would perceive it as being further to the left. . . . In crediting the rat with perceptions, one is already crediting it with beliefs. In perceiving that the food has moved to the left, the rat acquires the belief that the food has moved to the left. (1976, 164)

In addition to perception, memory is involved.

> The rat may also need to remember what it has perceived. If it loses sight of the food while swimming, yet continues to swim in the right direction, this is probably because it remembers that there is food on the opposite bank. Memories based on past perceptions are continuations of beliefs about the environment. (1976, 164-65)

Thus, plasticity theories, in explaining how an organism's behavior would change if certain circumstances were changed, presuppose that the organism has perceptions and beliefs and memories, that is, presuppose mental states.

In addition to claiming that goals are intentional objects, Woodfield claims that the means to reach a goal, as referred to in a teleological statement, is also an intentional object, but instead of being an object of desire, it is an object of belief. "I have characterized the content of the instrumental belief as 'that B is a means to G'" (1976, 168). As was the case with the intentional status of goals, the intentional status of means in the context of teleological statements has implied support even by plasticity theorists. This is clear from the fact that "if the rat wrongly believed that the food had moved further to the right, e.g., if two mirrors were so arranged as to make it look as if the food were further to the right, and started swimming to the right, we should certainly judge that behavior to be appropriate given the rat's belief" (Woodfield 1976, 165).

Woodfield selects as a standard way of describing goal-directed behavior the purposive teleological description, or purposive TD, "S does B in order to do G," and encapsules his ideas about the analysis being nonbehavioristic and goals being intentional objects in its analysans, "S does B because S believes (B $\Rightarrow$ G and G is good)" (1976, 206). "S" refers to a system, "B" refers to behavior, and "$\Rightarrow$" refers to the relation of causally contributing to. Although "G" suggests a goal, it does not refer to a goal, because that would inject an element of circularity into the analysis, reducing its explanatory value. "I take it as a condition on the adequacy of any of these kinds of description that they should mention G, but without containing the word 'goal'. They must be descriptions which reveal perspicuously what is meant when G is said to be a goal" (Woodfield 1976, 162).

Some, such as Nagel and Wright, claim that teleology cannot be analyzed in terms of intentional or mentalistic concepts because goal-directed behavior occurs also when conscious states cannot be assumed. Woodfield disagrees: "There is no need for an agent to be aware of the operation of his desire and belief when he acts purposively. He need not even know that he has them" (1976, 171). This observation applies not only to humans, but to animals, as well. "But it is well to remember that desires and beliefs need not be conscious. Since human beings can be in certain mental states without being aware that they are, it would be wrong to deny similar kinds of mental states to animals purely on the grounds that they are not aware of them" (Woodfield 1976, 172). Although it is unclear what can be said about animals, he is surely correct about humans. Much confusion has resulted from ignoring the large category of behavior, including, of course, much habitual behavior, which is intentional though unconscious and assuming erroneously that the mental is coextensive with the conscious.

If goals are intentional objects, the missing goal problem and the problem of impossible goals do not arise. Expressed in terms of a theory that takes goals as intentional objects, the missing goal problem pertains not to whether the intentional object is missing but whether the intentional object is realized extensionally. That may be a practical problem for the agent, but it is not a theoretical problem for the philosopher. "The core-concept of a goal is the concept of an intentional object of desire. No wonder, then, that the truth-value of a TD is unaffected by whether G occurs or not" (Woodfield 1976, 166).

Likewise, the problem of the proliferation of subgoals is resolved. Reaching a point one inch from the food is not a goal because it is not an object of desire. Where D is an earlier stage or a component of the sequence ending in G, it would be true that S did B in order to do D only if the cause of S's doing B were a combination of a desire to do D and a belief that B $\Rightarrow$ D. "But S might not have such a belief or desire, since S might not conceive D as a separate action" (Woodfield 1976, 170).

Woodfield's analysis of teleological statements about human artifacts is similar to his analysis of teleological statements about goal-directed behavior. However, instead of the more common "The function of X is Y" formulation, he chooses as his standard form "X does/has A in order to do G," which he calls an artifact function TD. Although with suitable adjustments, function statements can usually be cast into this form, it is awkward, for function talk normally uses the term "function" and is usually two-termed, whereas his standard form does not use the term "function" and is three-termed.

His analysis of "X does/has A in order to do G" is "X does/has A because S believes (A $\Rightarrow$ G and G is good)" (1976, 210). By means of the belief operator, artifact function TDs, like purposive TDs, refer to intentional objects. An arrow has feathers because the designer or maker believes that having feathers causally contributes to accurate flight and he desires accurate flight. Artifact functions are purposive, with the purpose determined not by the behavior of the artifact but by the intentions of S. "To settle what the purpose of an artefact is one must refer to the purpose of the designer or maker" (Woodfield 1976, 111). The sometimes vexing problem of how to allow for malfunctions vanishes, for the designer or maker may be fallible. The feathers do not need to be able to produce accurate flight to have the function of doing so, and this is so for the same reason that a goal need not be realized to be a goal. The function depends on the belief, not on the performance.

Traditionally, the most difficult area of teleology to analyze has been teleological statements that describe behavior, organs, and processes believed not directed or designed by mind. This usually is taken to include the overlapping areas of instinctive behavior of all organisms and all teleological behavior of lower organisms. It also includes the behavior and properties of parts and processes of all organisms, especially organs and their characteristic processes. The former, behavior of the whole organism without intentional features, Woodfield calls behavior functions. The latter, behavior and properties of the parts and processes, he calls biological functions.

Behavior functions play an especially large role since much of what others have called goal-directed behavior is not so regarded by Woodfield because of his mentalistic view of goals. He would not, for example, find goal-directed behavior nearly so widespread as does Nagel. The same behavior pattern, such as moving toward food, is directed to a goal if done by a chimpanzee but is not directed to a goal if done by a planarian. "The presence or absence of the psychological verb 'believes' marks a crucial dividing line"(Woodfield 1976, 211). The behavior of the planarian is still directed. However, instead of being directed to a goal, it is directed to an end. An end is not an intentional object but is simply something good.

The standard form version of the behavior function is "S does B in order to F." S can be an individual, a species, or a group. "G" has been replaced by "F." Although "F" might suggest a function, it actually indicates an end. In addition, the belief operator has been removed. The behavior function TD is analyzed as "S does B because B $\Rightarrow$ F and F is good" (Woodfield 1976, 206). An example is "Spiders spin webs in order to catch insects," which is analyzed as "Spiders spin webs because spinning webs causally contributes to catching insects and catching insects is good." This formulation correctly rejects the view that a function is merely an effect, a necessary condition, or a sufficient condition of something (Woodfield 1976, 114). "Good" refers not to desire but to what is beneficial to the organism or the species. By having the recipient be either, provision is made for behavior that does not help the agent and may, in some cases, be harmful to it. The term "benefit" covers survival and reproduction. "In fact, only two biological ends, survival and reproduction, will support the interpretations I have given" (Woodfield 1976, 139).

Behavior functions do not have intentional objects; hence, they do not presuppose mind. They are not purposive and are not in any way connected to goals. Behavior functions are not particular actions but are, rather, kinds of actions. The behavior is always characteristic behavior, behavior which that type of organism characteristically does in those circumstances (Woodfield 1976, 209). The instinctive scattering of chicks for cover when a hawklike shadow passes over is characteristic behavior for chicks, while a chick's remaining in place is not.

As noted above, the second area of teleology that has been traditionally difficult to analyze is that of teleological descriptions of the parts and processes of organisms, especially organs, their structure, and characteristic behavior. This fourth and last category Woodfield calls biological functions. Whereas behavior functions are functions of the whole organism, biological functions are functions of components of organisms. The biological function TD is "X does/has A in order to do F." Its analysans is "X does/has A because A $\Rightarrow$ F and F is good." As with the case of behavior functions, there is no intentional operator, so a mind or a designer is not presupposed. A and F are types or kinds, usually of organs. "A $\Rightarrow$ F" means that A characteristically and normally contributes to F. The traditional example of the heart fits here. "The heart beats in order to circulate the blood" is analyzed as "The heart beats because beating contributes to circulation of the blood and circulation of the blood is good."

Summarizing Woodfield's four-part analysis, we have:

Purposive TD: "S does B in order to do G"
  is analyzed as
"S does B because S believes (B $\Rightarrow$ G and G is good)."

Artifact function TD: "X does/has A in order to do G"
  is analyzed as
"X does/has A because S believes (A $\Rightarrow$ G and G is good)."

Behavior function TD: "S does B in order to do F"
  is analyzed as
"S does B because B $\Rightarrow$ F and F is good."

Biological function TD: "X does/has A in order to do F"
  is analyzed as
"X does/has A because A $\Rightarrow$ F and F is good." (Woodfield 1976, 206)

He regards this list as complete. "I do not think, however, that there are any genuine alternative schemata" (1976, 213). The key parts are a causal component symbolized by "$\Rightarrow$," an evaluative component symbolized by "good," and an overall explanatory role, conveyed by "because." "The causal clause identifies an actual or envisaged effect of a certain event, the evaluative clause says that this effect is good from some point of view, and the whole TD says that the combination of these elements provides *raison d'etre* of the event" (Woodfield 1976, 206).

There is a general impression that there is some underlying unity that makes teleology a single subject and is responsible for the fact that all teleological statements can be expressed in the means-end language of "in order to." Woodfield sees this unifying factor as being exhibited in his four standard form TDs: "The point of fitting all TDs into a single general table with headings for the separate sentence-components is to show that they have structural similarities despite their logical diversity. . . . The matrix reveals, however, that the various TDs are unified at an abstract level" (1976, 212-13).

The only common properties are the occurrence of the words "because" and "good." "Because" indicates that teleological statements are explanatory, but that is not enough to ground our impression that teleology is a single subject. That leaves only the occurrence of the word "good." It is this, apparently, that Woodfield sees as the unifying factor. The table of four types of TDs, he says, "enables one to see that teleological explanations have a sort of unity. The fact that the explanandum event has, or is regarded as having, a good effect is presented as a reason for its occurrence" (1976, 212-13). "Their structural similarities are striking enough, I think, to justify the conclusion that the different types of teleological explanations are variations on a single theme. This theme is, to put it simply, the idea of a thing's happening *because it is good*" (Woodfield 1976, 205).

Unfortunately, the unity of the analysis is weakened when Woodfield acknowledges, "More exactly, the TDs I deal with convey the idea that the thing happens or exists because it leads or is believed to lead to something which is good" (1976, 205). There is a gulf between something actually leading to something good and the mere belief that it does. In fact, the four schemata, themselves, reveal that gulf because the analysans of the behavior function TD and the biological function TD say that something is good while the analysans of the purposive TD and the artifact function TD say only that something is believed to be good. Since a thing can be believed to lead to something good when it does not, Woodfield has inadvertently removed the idea of a thing's happening because it is good as the feature common to the four types of analyses.

Nevertheless, there remains, one would think, a measure of unity since the term "good" presumably has one meaning in all four schemata. This presumption is the next to go. Regarding the biological function TD, Woodfield says, "'F is good' is understood to mean that F is good for S (in normal circumstances), either intrinsically or because it characteristically contributes to some further good" (1976, 208), and, regarding the behavior function TD, he says, "It is understood that F is good *for S*, either *qua* individual agent, or *qua* species, in case F promotes reproduction, survival of offspring or of conspecifics" (1976, 209). Speaking of both, he says, "If the end is actually good for a being, it is what I call a natural or biological end" (1976, 211) and "In fact, only two biological ends, survival and reproduction, will support the interpretations I have given" (1976, 139). Thus, for the analyses of behavioral function and biological function TDs, "good" means good for S or beneficial to S, taking S as either the individual or the species, with what is good or beneficial limited to survival or reproduction.

However, "good" does not mean good for S or beneficial to S in purposive TDs and artifact function TDs. Regarding purposive TDs, he says, "'G is good' is interpreted as meaning that G is good to do, so that the belief that G is good amounts to a desire to do G" (1976, 207). Regarding both purposive and artifact function TDs, he says, "All that is meant here by 'S believes that G would be good to do' is that S wants to do G" and "I propose, therefore, to go ahead and substitute 'S believes that G is good' for the clause 'S wants to do G' in my analysans" (1976, 204). Thus, the unifying element in Woodfield's analysis of teleology disintegrates, for, in the analyses of two kinds of TDs, "good" means *beneficial*, and in the analyses of the other two, it means *desired*. What is desired may, of course, not be beneficial at all—a point he acknowledges: "Nor do I believe that we always desire the good" (1976, 204).

In addition to the fact that Woodfield has the term "good" mean two quite different things, thereby destroying the common element in his four explanation-schemata, a further problem is that his use of "good" causes his purposive TD and artifact function TD to claim that all goals are objects of desire. Indeed, he asserts it explicitly: "I stick by my arguments for the thesis that having G as a goal (in the 'core' sense) involves being in the mental state of desiring G" (1976, 204).

Woodfield is certainly with good company in connecting goals to desire. Ducasse, in the article referred to earlier, wrote that act X is purposive only if there is "*Desire* by the performer that *Y* shall occur" (1925, 154). Although Alvin Goldman allows intermediate actions and goals to be not desired, as waving one's hands above one's head, he requires that the ultimate action or goal be desired, as attracting someone's attention. "*Act-token* A *is intentional only if there is an act(-type) A' such that the agent S wanted to do (exemplify) A'''* (Goldman 1970, 51). "What, then, is implied by saying that *S* flipped the switch 'in order to' turn on the light? Evidently, this explanation implies that *S wanted* to turn on the light" (Goldman 1970, 78).

Normally, when one's goal is food, one desires food; when one's goal is rest, one desires rest, etc. However, such is not always the case. Some goals are not desired, and it is quite possible to do a thing in a goal-directed manner that one does not wish to do. A child, to his acute embarrassment and much against his will, may be required to return to the store the toy he stole. The action is certainly directed to a goal, for he could explain, if asked, why he is carrying the toy to the store. If the action were desired, as Woodfield alleges, it would make incomprehensible why he endeavored to avoid or postpone the assignment, why he looked downcast, why he walked up and down the sidewalk in front of the store before entering, and, especially, why he said he did not want to do it. Indeed, the painful memory, we might suppose, lasted a lifetime and modified his adult behavior. To say he performed the action because he desired to escape punishment, even if true, would not render false the statement that he did that which he did not desire to do, for it is the latter statement, not the former, that explains his avoidance behavior, his demeanor, and his modified later life.

Indeed, Woodfield notes a similar case: "A soldier may intend to carry out an order that he wishes he did not have to carry out. In one familiar sense of 'want', he does not want to carry it out" (1976, 202-203). This much is indisputable. However, he then adds, "But in the sense in which I use the word (which is also familiar), he does want to, all things considered, in so far as he intends to" (1976, 203).

Although dictionaries do not settle philosophical issues, they do register standard usage. Neither standard desk dictionaries nor the *Oxford English Dictionary* recognizes Woodfield's allegedly familiar meaning. They do not define "intention" in terms of desire or refer to desire in their definitions in any way whatever.

The argument can be put another way by noting that signs indicating aversion in respect to an action that is not performed may be present in respect to an action that is performed. If those signs indicate aversion in the former, they should in the latter, as well. Consider Jones. He says he does not want to pay his taxes. When the subject of taxes is raised, he frowns and speaks with irritation. He votes against candidates who propose increasing taxes. He goes to considerable trouble to locate tax shelters. Finally, he does not pay his taxes. Now consider Smith. He also says he does not want to pay his taxes. When the subject of taxes is raised, he also frowns and speaks with irritation, votes against candidates who propose increasing taxes, and goes to considerable trouble to locate tax shelters. However, unlike Jones, he does pay his taxes. If one judges that Jones does not desire to pay his taxes and is then asked to justify that statement, one would certainly cite as indicators the fact that Jones says he does not want to pay his taxes, that he gets irritated when the subject is raised, and so on. However, those are characteristics of Smith's behavior, as well. What are signs of aversion in one should be signs of aversion in the other. It is true that in the case of Jones, one would also refer to the fact that he did not pay his taxes, but that would be only mentioning one indicator among several. If Jones's nonpayment of taxes were the only aversion indicator, the others would not have been mentioned.

Woodfield is, of course, aware of such cases; indeed, his example of the reluctant soldier is such a case, for he says that one "may regard doing G . . . as the least unpleasant of the alternatives open to him" (1976, 203), but he disregards the import for his theory. Just as one thing can be less tall than another without ceasing to be tall, so can one thing be less repugnant than another without ceasing to be repugnant. With the house on fire, a devoted mother may have time to rescue only one of her two children, and so must choose which to abandon. To say that she really desires abandoning one of her children is going to unreasonable lengths to protect a theory. The fact that she would take umbrage at such a statement indicates that she would regard it as outrageously false, and she ought to know. Perhaps a psychosis results from the terrible choice. The examining psychiatrist would never say that the mother desired abandoning her child; only a philosopher would make such a statement and only other philosophers would give it credence.

One can also argue from a feature of ethics that intentional actions need not be desired. A person may be obligated to do an action he does not want to do. It is also widely, though not universally, held that one can be obligated to do only what one is able to do, that "ought" implies "can." However, if one may be obligated to do what one does not desire to do and if obligation presupposes ability, then there must be some actions one is able to do that one does not desire to do, some action one can do while lacking the desire. Being able unconditionally to do something does not mean being able to do it if one desires to do it, for that is only conditional ability; it means being able to do it whether or not one desires to do it.

A further problem with Woodfield's analysis is that if the purposive and artifact function TDs are interpreted as he directs, they become incoherent. The analysans of the purposive TD is "S believes (B $\Rightarrow$ G and G is good)" and the analysans of the artifact function TD is "S believes (A $\Rightarrow$ G and G is good)." The part "S believes (. . . G is good)" occurs in both and is interpreted as meaning S desires G. Such an interpretation could, of course, be resisted on the ground that the words employed just do not have the required meanings, but Woodfield avoids this criticism by saying, "I am, if you like, stipulating a use for the word 'good' in this special context" (1976, 204).

However, if "S believes (. . . G is good)" means S desires G, one would think that "good" and "desired" must be taken as synonyms. If they are synonyms, we should be able to rewrite "S believes (. . . G is good)" as "S believes (. . . G is desired)." Such, however, cannot be correct, for Woodfield means not that S believes that G is desired but that S desires G. Alternatively, we might try "S believes (. . . S desires G)." This also cannot be what he wants, for he means not that S believes that S desires G (whatever that may mean), but simply that S desires G. A more radical reformulation would be "S believes (. . .) and S desires G." This version captures the correct sense, for one part of Woodfield's meaning is that S believes that B (or A) causally contributes to G, and the other part is that S wants or desires G. However, this formulation does not fit the schema Woodfield provides. The one he provides is "S believes (. . . G is good)," in which "G is good" falls under the belief operator. This schema is also the one he needs, for without it, his claim to have analyzed teleology in a way that preserves the unity among teleological statements would not have even superficial plausibility. Thus, not merely does "good" mean desired only if it is so stipulated, but the newly interpreted term cannot be reinserted in the required schema. That is, if "good" and "desired" have the same meaning, then "S believes (. . . G is good)" is incoherent, which arises from the persistent fact that desires are not beliefs.

One might wonder, however, if these remarks do not arise from a faulty premise. Perhaps the claim that "S believes (. . . G is good)" means S desires G implies not that "good" means desired but rather that "believe good" means desired. To clarify his meaning, Woodfield suggests looking at the expression "see as good." "Perhaps the expression 'see as good' is better than 'believe good'" (1976, 204). This does not help, however, for to see something as good is merely to take a perspective that emphasizes favorable features. To see a natural disaster, such as a flood, as good is to call attention to such things as improved social cohesion resulting from it. To see the Roman wars of conquest as good is to emphasize such benefits as the Pax Romana, which followed. The good that is emphasized is, no doubt, usually also desired, but, as Woodfield notes, not always. Seeing something as good is, therefore, not the same as desiring it. The distance between facts and desires is as great as that between beliefs and desires.

Further, Woodfield alters the concept of belief so that it no longer can be in error. "Another qualification concerns the word 'believe'. It denotes an attitude which is too cognitive. . . . The person who believes that G would be a good thing to do, in the sense I am after, need not feel that there is a right answer, which someone else could tell him perhaps, about whether G really is a good thing to do" (1976, 204). Since, in the analysis of the purposive TD, "S believes (B $\Rightarrow$ G and G is good)," one belief operator covers both conjuncts, what Woodfield says about the belief that G is good must apply also to the belief that B causally contributes to G. This interpretation implies that an agent's belief, say, that pouring water on fire will causally contribute to extinguishing the fire, is a belief with such a nature that one "need not feel that there is a right answer, which someone else could tell him perhaps, about whether" pouring water on a fire really does causally contribute to extinguishing it. Expressing it more simply, the position is that the agent's belief that water puts out fire does not claim that water puts out fire. Such a view of the nature of belief is incorrect and makes the concept of belief unintelligible. Similar observations apply to the analysis of the artifact function TD, "S believes (A $\Rightarrow$ G and G is good)."

The "because" in the analysans indicates an explanation, presumably a causal explanation, for all four types of TD and must be a causal explanation in the case of behavior and biological functions, since their analyses have no recourse to anything mental. Therefore, although it would not be acceptable to say, "Animals have stomachs in order to circulate the blood," it would be acceptable to affirm the analysans, "Animals have stomachs because stomachs causally contribute to circulating the blood and circulating the blood is good." This is so because among the many long-range consequences that having a stomach causally contributes to, one is that of circulating the blood. Similarly, "Animals have lungs in order to see" is false, but its analysans "Animals have lungs because lungs causally contribute to seeing and seeing is good" is true, and "Animals have kidneys in order to hear" is false, but its analysans "Animals have kidneys because kidneys causally contribute to hearing and hearing is good" is true. This class of counterexamples is based on the fact that a remote effect can be beneficial without being a function.

Although he regards his four TDs and their respective analyses as complete, saying, "I think that every true 'in order to' sentence is equivalent in meaning to a sentence exemplifying one or other of these schemata, or can be assimilated to a standard form that is" (1976, 210), Woodfield offers, in effect, a fifth analysis. This analysis talks about a metaphorical extension of a purposive TD: "There is a core-concept and a broadened concept, such that the core concept of having G as a goal involves the concept of wanting G, and such that the broadened concept involves the concept of either wanting G or being in an internal state analogous to wanting G" (1976, 163-64). It has application not only to lower organisms, but also to machines. "S may be an artefact, like a robot. Assuming that such systems do not literally have desires and beliefs, TDs of them would not be objective statements, according to this analysis" (1976, 207).

Though originally metaphorical, Woodfield, like Wright, indicates that the metaphor may now be dead and that such teleological statements now have literal meaning. "However, it is not realistic to say that purposive or goal TDs are always metaphorical in such cases. They may have been figurative originally, but many people would say that the metaphor is now dead" (Woodfield 1976, 207). The intentional operator "believes" still occurs as part of the analysis, but may be now read as a dead metaphor and referring to a quasi-belief, belief-analogue, or belief-like state. "I suggested that a TD like 'The robot turned left in order$_G$ to get back to base' should be interpreted to mean that the robot turned left because the robot 'believed' (quasi) that (turning left would contribute to getting back to base and getting back to base would be a good thing to do)" (Woodfield 1976, 207).

Metaphorical talk is, of course, common. Gardeners talk about ferns "liking" shade or blueberries "preferring" sandy soil and physicists talk about a particle "feeling" the force of a magnetic field and of plasma "trying" to escape magnetic confinement. Such openly metaphorical language raises no philosophical issues peculiar to teleology. However, were "believe" in "S believes (B $\Rightarrow$ G and G is good)" used as an ordinary metaphor, any claim to have given a philosophically interesting analysis would vanish. If saying, "The spider built a web in order to catch insects" is to be understood the way we understand "The sun smiled on the landscape," teleological statements in the life sciences would be uncontroversial and no one would bother either attacking or defending them.

The situation changes, however, when Woodfield says that the metaphor is sometimes dead. As noted in the discussion of Wright, a dead metaphor is no longer a metaphor but, rather, an expression that formerly was a metaphor. It is now an expression with a quite literal meaning and is as definable as other literal expressions. Dictionaries concur, and define "wolf" as "to eat greedily," "cap" as "to cover the end of," "thumbnail" as "very small or brief," and "full-fledged" as "total or complete"—all dead metaphors, all formerly metaphors but metaphors no longer. If the claim is that the original meaning of purposive TDs is the one appropriate to humans and higher animals and that this meaning has since been extended by dead metaphor to lower animals, plants, and machines, there must now be a new literal meaning and a definition. If "believes" has such a new defined meaning, it should have been given. Where the meaning remains too elusive to be confined to a definition, the metaphor is not dead.

Woodfield states,

> I shall not list a separate analysans for purposive TDs of inanimate systems, but shall leave it to be understood that S's internal state can be either a belief or a belief-analogue. Equally, then I must allow that purposive behaviour can occur among living systems that don't have fully-fledged beliefs, but only belief-like states. (1976, 207)

This means, then, that he has two analyses of goal-directed behavior of lower organisms. One is that given above in his analysis of the behavior function TD in which "S does B in order to do F" is analyzed as "S does B because B $\Rightarrow$ F and F is good." Here there is no reference to belief. The other is a metaphorical version of his analysis of the purposive TD, "S does B in order to do G," which states that "S does B because S believes (B $\Rightarrow$ G and G is good)." There is reference, not to belief, but to something that in some unspecified way more or less resembles belief. These two analyses of the same behavior are very different from each other. They are competing, not complementary. Furthermore, since similarity has no natural boundaries, an analysis in terms of belief-like states excludes nothing. Although Woodfield wants to include simple organic life and robots, there is nothing that would exclude the behavior of rock slides and rain storms or the movement of tectonic plates or galaxies. The fact that Woodfield goes on to conjecture that future robots may be able to rejoice, do tasks willingly, experience filial gratitude, prize or value things, all apparently in literal ways, is unsettling and is hard to reconcile with his earlier careful work on the nature of goals (Woodfield 1976, 222).

One of the problems every analysis of teleology must face is how to avoid implying reverse causation. This is not a problem for Woodfield's purposive TD and artifact function TD because they utilize a belief operator. The analyses of these TDs do not presuppose that later events cause earlier events because it is the beliefs about these events, not the events, themselves, that are appealed to in explaining later events, and the beliefs occur prior to the events in question. This is not the case, however, with the behavior function TD and the biological function TD. An example of the former, borrowed from Woodfield, is "Birds sit on their eggs in order to hatch their young," which is analyzed as "Birds sit on their eggs because sitting on their eggs contributes to hatching their young and hatching their young is good." In this context, "good" means beneficial to the individual or the species (Woodfield 1976, 209). Since the benefits of hatching the young, which would be the life of the new generation and its reproduction, happen after the parents' egg-sitting, the analysis seems to imply that later events cause earlier ones. An example of a biological function TD, also used by Woodfield, is "Hearts beat in order to circulate the blood." This is analyzed as "Hearts beat because beating contributes to circulation of the blood and circulation of the blood is good." Since circulation follows beating and the benefits follow circulation, this seems to imply that the heart beats because of the results of the beating, thus, that later events cause earlier events.

Woodfield's solution, like Wright's, is to say that function statements are about types, not individuals. About biological functions, he says,

> "A ⇒ F" means "A characteristically and normally contributes to F", where A and F are types. . . . Standardly, the TD is a tenseless general statement about *types* of organs. "The heart beats in order to circulate the blood" is about hearts in general, of whatever individual of whatever species; "Peacocks' tail-feathers have bright colours in order to attract peahens" is about the tail-feathers of peacocks in general. (1976, 208)

In another place, he says,

> The functional TD is not referring to any particular heart, but to hearts in general, or to the heart, *qua type* of organ. Consequently, it is not about any particular occasion of beating, but about heart-beating in general. It asserts "Hearts in general beat because beating

(characteristically) contributes to blood circulation, and blood circulation
is good for the organism." (1976, 137)

He calls this the phylogenetic interpretation. Since he says that "'S does
B' is a general statement in the behaviour-function schema, a singular
statement (normally) in the purposive schema" (1976, 210), this
interpretation has the unlikely consequence that a higher organism can
have behavior directed to particulars but a lower organism cannot, that a
human can pick an individual egg from a nest in order get food but, using
his example above, a bird cannot sit on an individual egg to hatch young.

As indicated earlier in the examination of Wright's and Neander's use
of this strategy, moving from describing this X to describing Xs in
general or from describing this heartbeat to heartbeats in general is
generalization only. Other operations are not performed as well. If
statements about kinds and types did not express the same causal
sequences as the statements about the individuals of which they are the
generalization, generalizations about causal matters would be of little
value. A physician reading in his desk manual that low iron in the blood
causes fatigue would be unable to determine, regarding the patient before
him who exhibits both low iron and fatigue but possesses only individual
organs and individual organic processes, whether his low iron caused his
fatigue or his fatigue caused his low iron and, consequently, would be
unable to determine whether to prescribe iron medication or bed rest. As
a matter of fact, the physician has no such problem. Thus, although
Woodfield rejects reverse causation, saying, "Teleological explanation
cannot possibly be committed to any reversed causal hypothesis" (1976,
34), his analysis, like Wright's, implies it.

Another difficulty with the type solution to the reverse causation
problem is that we do make teleological statements about individuals, as
well. Woodfield agrees: "But it is possible to ask functional 'Why?'
questions about individual specimens" (1976, 208). How can the
implication of reverse causation be prevented? He explains by giving
what he calls an ontogenetic explanation. "Furthermore, it is possible to
give an 'ontogenetic' functional explanation of why a particular heart
beats. The TD would say that this heart beats because its own past
beating has benefited the owner by helping him to survive" (1976, 208-
209). He makes similar remarks about interpreting the behavior function
TD. Thus, he offers an alternative interpretation of teleological
descriptions of all organs and internal processes and all instinctive
behavior and behavior of lower life. An alternative analysis of such a
large class of teleological descriptions should have been developed fully

and given a place in his formal table of analysis. However, such an analysis comes at high cost, for to say that this heart beats because its past beating benefited the owner does not differ in form from an explanation that says, "This heart beats because of the electrical stimulation it receives" or "This heart beats because the blood supplies it with glucose." Such formulations remove the forward orientation that distinguishes teleological statements from ordinary causal statements. If that were acceptable, the rest of Woodfield's analysis would not be needed.

In discussing functions, he makes the interesting statement, "Admittedly, functional explanations convey the *impression* that later events are influencing earlier ones, but this is an illusion based on the fact that they fudge tense-differences" (1976, 138). Later, the idea is repeated: "Functional TDs sum up a number of historical facts by fudging time-references, thereby creating the illusion that the cause of the present beating is the fact that it will have a beneficial effect" (1976, 209). It is surprising that this candid observation did not prompt a reassessment of the steps that led to it, for if function statements fudge facts and create illusions, it makes inexplicable Woodfield's defense of function statements in the life sciences, which certainly do not aim at fudging facts or creating illusions.

Woodfield's analysis of behavior function does cover large areas of instinctive behavior and behavior of lower organisms. However, like Wright's consequence-etiological analysis, the causal claim, in this case expressed by "⇒", makes it vulnerable to examples of behavior directed to impossible ends. This is so despite the fact that Woodfield, like Wright, is aware that an acceptable analysis of teleology must make provision for ends and goals that cannot be achieved. An example discussed by both, that of a fish caught in a net, illustrates the problem. If the net is sturdy and the mesh small, escape is impossible. Although the behavior TD "The fish swam against the net in order to escape" is true, the analysans "The fish swam against the net because swimming against the net causally contributes to escape and escape is good" is false. Any response that introduces a probability of escape, however slight, can always be met by modifying the example to reinstate complete and utter impossibility. Restrictions about types and kinds can be met by talking about the class of fish confined in strong nets with small mesh, and swimming, of course, meets the requirement that the behavior must be a characteristic kind. Other examples can easily be added.

A slightly different class of end-directed examples can be constructed around injured organisms. We do not restrict our teleological descriptions to normal healthy specimens. In the competition for survival, organisms are often so severely defective or injured that their behavior, although still characteristic, has no chance, whatever, of success. In those cases, the behavioral function TD and biological function TD are true, but the analysans of each, committed to the claim that the behavior must causally contribute to F, is false.

Yet another class of behavior directed toward impossible ends can be found in the radically unbalanced struggles in nature, as in the pursuit of the slow, small, and weak by the swift, large, and strong. If one selects a narrow enough class of such instances, such as beetles pursued by owls in an area of no cover, the teleological description becomes true and the analysans false, for the beetles run in order to escape even though that running does not causally contribute to escape. We regard such classes of examples as irrelevant to the question of whether the behavior is teleological, although they are clearly not irrelevant to Woodfield's analysis.

Because Woodfield says that the behavior normally contributes to F, exceptions can be found also in instances in which success is possible but improbable. Cases in which predators usually win fit in here, such as the predation of young sea turtles in their race to the sea. We say they run to escape the predators, even though that behavior normally fails, making the teleological statement true and the analysans false.

One might feel, however, that the above examples are inappropriate, that when Woodfield says the behavior must be characteristic behavior, he means that the behavior talked about in the analysans must be very broad and general, for example, not running to the sea or running from gulls, but just running from pursuit. It is that kind of action that is to be considered characteristic action and that is normally successful. Of course, swimming through fine strong mesh is impossible for adult tuna, but the behavior of swimming through strong mesh is not the level of behavior that teleological statements talk about. The more general behavior of just swimming usually does have beneficial results.

There is something correct about this line of thought. If teleological language is ever to be properly analyzed, it seems certain that the level of generality of actions and properties will play a role. Nevertheless, we do make teleological statements about very detailed and specific behavior. We say about this particular trout that it is now swimming against the sides of this particular net in order to escape or that this particular moth is fluttering against that window pane in order to get out, but analyzing these statements according to Woodfield's directions yields false statements.

Another class of counterexamples to Woodfield's analysis of behavioral function TDs involves looking for cases in which the end pursued is not good for either the organism or the species, as in the case of a lower organism eating a poisonous substance. The behavioral function TD, "The ants entered the box in order to eat the bait," is true, but the analysans, "The ants entered the box because entering the box causally contributed to eating the bait and eating the bait is good," is false since eating the bait is not good either for the individual or the species. One cannot, of course say that such an example is inapplicable because the ants made a mistake and thought they were eating food, since the analysis admits neither norms nor anything mental.

A further point concerns Woodfield's explanation of the variable "G." "'F' and 'G', however, are always names for act-*types*" (Woodfield 1976, 210). It is patently false that goals pursued are always types or kinds. That wolf is not chasing deer in general, but that one lagging behind. One rummages around looking not for keys in general but for a particular individual key to the ignition of a particular automobile. Even searching for several keys does not involve a goal that is a type or kind but a plurality only. In addition, goals need not be actions at all; the above are not.

Analysts writing on teleology have differed on the issue of whether teleological statements are inherently explanatory. Woodfield holds a curiously mixed position that some are not but that others are, specifically, that function statements, that is, statements using the word "function," are not inherently explanatory, but that functional TDs, which do not use the word "function," are. An example of what he calls a function statement is "The function of the heart is to circulate the blood." The corresponding functional TD is "The heart beats in order to circulate the blood." The former has the form "The function of X is F" and the latter has the form "X does A in order to do F." To support his position, he cites the Hawthorne study of factory workers that had the unexpected result of increasing production, something called a latent function (1976, 109). He notes that the study was not done in order to increase production, making the functional TD, "The investigators performed the study in order to increase production," false. To make Woodfield's case, one must, as Woodfield does, take the corresponding function statement, "The function of the study was to increase production," as true. "Latent function," however, is a technical term in sociology. Something may be a latent function without being a function in the sense that hearts have a function. There are many different uses of the term "function," including the mathematical use, something "functioning as" something else, something "functioning well," etc. "Latent function" is one of the variants. Developing a coherent analysis of functions requires keeping distinct uses separate.

A second example Woodfield uses to support his case that function statements do not explain while functional TDs do is that of rainfall. "Again, it is certainly not true that rain falls in order to water the crops; but to say that rain performs the function of watering the crops seems to me not *false*, but misleading" (1976, 109). On the contrary, the statement "The function of rain is to water the crops" not only is false, it has the obvious kind of falsity that authors find useful to counter the view that functions are beneficial effects, along with such examples as "The function of corn is to provide food for animals" and "The function of the nose is to hold up eyeglasses." If one reads "Rain performs the function of watering the crops" as true, one must interpret it differently from "The function of rain is to water the crops." This can be done if a suitable context is supplied, as, for example, that of lecturing to a class on crop production in which the method of presentation is in terms of seeing what contributes to the goal of a good harvest. Certain conditions contribute to reaching that goal, such as fertile soil, sufficient drainage, effective weed control, moderate fertilizer costs, etc., and in that context one might also list such things as adequate sunlight and adequate rainfall.

If Woodfield were correct in his claim that function statements, such as "The function of the heart is to circulate the blood," are not explanatory while functional TDs, such as "Hearts beat in order to circulate the blood," are, then they could not have the same meaning. However, if function statements had different meaning from his functional TDs, his analysis of function statements would be threatened, for the point of his offering his artificial functional TD, behavioral functional TD, and biological functional TD was that function language could be assimilated into one of their three forms.

Woodfield's analysis is the most complex one in the literature, but one of the most interesting. It does much better in respect to purposive TDs and artifact function TDs, where mental states are admitted, than with behavior function TDs and biological function TDs, where they are not. It is interesting that it does not become tortuous and hard to follow or invite easy counterexamples until mental operations are removed.

Chapter 7

Recent Applications

The problems of how to understand goal-directed behavior so that reverse causation is not inadvertently assumed or how to understand goal-directed behavior when the goal is not realized or when the behavior is consistent with several goals all point to the need to include something about representation in the analysis. The parallel problem of an item's having as a function what it cannot produce, as a knife having a function to cut when it cannot cut, likewise points to the need for construing functions as somehow involving representation. We have seen that not making provision for representation comes at high cost, such as equivocation on the concept of goal, disallowing teleological language about individuals, ascribing causal power to abstract entities while denying it to particulars, accepting bizarre and transient collections as legitimate teleological systems, and more. If anything is to be learned from all this, it surely is that teleology involves representation and involves it essentially. It is, therefore, of some interest to observe in the last few years language theorists turning to the concept of biological functions to understand representation.

The interest is understandable. A naturalistic theory of language must be expressed in terms acceptable to natural science, and that seems restricted to what is broadly observable by the senses. Observation, however, is always of what is. It is not of what should be. The fact that questions regarding the foundations of ethics have exercised thinkers since ancient times, with no end in sight, points to the extreme difficulty in generating how things should or ought to be from how things are. Since language devices, especially those that accommodate misrepresentation and error, must be understood in terms of what they are supposed to accomplish, that is, in terms of norms, a problem close to the central problem of ethics confronts the philosopher constructing a theory of language.

Millikan

Among the recent theories of language utilizing the concept of biological functions is the much discussed theory of Ruth Millikan. The task, she says, is to define "designed to" and "supposed to" in a naturalistic, nonnormative, and nonmysterious way (Millikan 1984, 17). She accomplishes this by defining key concepts broadly enough to encompass both biological and linguistic phenomena. The central concept is that of proper function, the fundamental variety of which is the direct proper function.

> Where m is a member of a reproductively established family R and R has the reproductively established or Normal character C, m has the function F as a direct proper function iff: (1) Certain ancestors of m performed F. (2) In part because there existed a direct causal connection between having the character C and performance of the function F in the case of these ancestors of m, C correlated positively with F over a certain set of items S which included these ancestors and other things not having C. (3) One among the legitimate explanations that can be given of the fact that m exists makes reference to the fact that C correlated positively with F over S, either directly causing reproduction of m or explaining why R was proliferated and hence why m exists. (Millikan 1984, 28)

Direct proper functions are determined by having a certain kind of past, not by having a certain current ability or disposition. "According to my definition, whether a thing has a proper function depends on whether it has the right sort of history" (Millikan 1989[c], 292). Although there is more emphasis on past performance, the definition shows a marked resemblance to Wright's etiological analysis. Statement (2), above, is similar to Wright's "Z is a consequence (result) of X's being there," expressed as differential correlation, (1) sets that correlation in the past, and (3) says about what Wright says in "X is there because it does (results in) Z."

Because current performance or capability plays no role, the definition correctly provides for malformed and damaged organs retaining their functions. It also correctly excludes side effects, such as heart sounds, since C must causally contribute to F whereas heart sounds do not contribute to circulation. Items Millikan lists as having direct proper functions include organs, inherited behavior, learned behavior, reproduced artifacts, and a great variety of linguistic devices (Millikan 1984, 28-29). However, her definition, though broad and carefully crafted, has several unfortunate consequences.

The definition requires direct proper functions to have three parts: m, the item having the function; C, which appears to be the mechanism or means for producing the function; and, F, the function. It, therefore, is appropriate only for those descriptions of functions that refer to some mechanism producing the function. In most cases, this is lacking, and is lacking even in Millikan's own examples, such as that of the heart's having the function of circulating the blood. Since m is that which has the function, it must stand for the heart, not the organism that has the heart. F, of course, is the function. The heart does its job by some means or other, but it is usually not specified, nor does Millikan do so. Another frequently used example is that of the chameleon. "The chameleon's pigment-rearranging device has a relational proper function. It is supposed to produce a color for the chameleon that bears the relation 'same color as' to the chameleon's nether environment" (Millikan 1984, 39). There is no reference to the means by which the device produces the color transformation, nothing for C to do. An example that figures especially prominently in her work is that of the bee dance. "The device in the bee that is responsible for this dance has a proper function to produce movements that bear a certain relation to . . . the direction (relative to the sun), distance, quality, and/or quantity of the nectar spotted" (1984, 39). There is that which has the function, the device in the bee dance, and there is the function—that is all. Again, there is nothing for C to do.

Apparently, when a function is understood as a two-part relation, it has been decomposed as far as our interest goes. Even if causal intermediaries were identified, and they are frequently simply not known, they would, in turn, have their own functions. We can say, of course, that the heart performs its function of circulating the blood by periodic contractions, but these contractions have their own function, to eject blood under pressure into the arteries. Presumably there is some means corresponding to C by which this is done, something about the chemistry of smooth muscle contraction, but whatever it is, it is simply not required for a legitimate function claim. Furthermore, sooner or later, since an infinite series of mechanisms must be rejected, we must come to an item having a function with no mechanism. Is the function of a hammer to pound things or is it rather to pound things by means of . . . what? The head? Perhaps, but it seems inappropriate to require in the primary definition of function an element that does not show up in informed function talk and usually is not known to or is ignored by the accomplished speaker. Of course, Millikan may be prescribing, not describing, but that would weaken her claim that she is getting norms from existing features of biology.

We do not ascribe functions to all intermediate stages and events. Millikan's analysis does. We do not say that a function of the heart is to push eosinophils through the first centimeter of the left subclavian artery, yet the history of the heart doing this is the same as the history of the heart circulating the blood, and the definition of proper functions fits both. If one thinks the heart does have such a first centimeter subclavian artery function, finer grained causal intermediaries can always be found, limited only by current knowledge of biochemistry. There is always a level of detail where function talk no longer applies. Her analysis, however, has no such restriction. In her elaboration of kinds of functions, she divides the bee dance up into many stages, but it is interesting that they are all stages where one could also speak of purpose.

Her definition requires that all sorts of effects that circulation contributes to are also functions of the heart. Since circulation contributes to everything the body does, this would make it a function of the heart to digest food, another to see, another to walk, and so on. She seems to accept this awkward consequence, calling them "serial functions," even offering the example of utterance as a function of the heart (1984, 35). However, since the definition does not block specificity and detail of effect, it also must include as other functions of the heart hearing rustling oak leaves, seeing storm clouds, tasting springwater, producing antibodies for scarlet fever, and healing scratches from rosebushes. Adding time and place parameters yields even greater specificity. Further, since the other organs and parts and processes of the body also contribute to these same things, they have the same functions, making it, for example, a function of the femur to produce endorphins. These consequences would make the functions of the heart innumerable and uninteresting and function talk quite useless. Again, one can view her definition as stipulating how functions are to be understood within this theory of language, but only at the cost of distancing that theory from the traditional theory of biological functions, from which she wishes to extract the needed norms.

Real world functions have a generality that is completely lost in this analysis. The function of the eyes of rabbits is to see, not to see longhaired dogs or trees between three and four inches in diameter, yet the history is the same and the definition fits both.

The definition says that certain ancestors of m performed F but says nothing about the number of ancestors or generations or length of time needed. This makes it too inclusive to serve as a sufficient condition for items to acquire functions by natural selection, such as Millikan's examples of the heart or inherited behavior. The absence of a minimum in ancestors, generations, or time is, of course, needed for the definition to be applicable to learned behavior, which often occurs in one trial and may occur in an instant.

By defining proper functions in terms of history, everything lacking the history is excluded. If an exact duplicate of an individual human accidentally and spontaneously came into being, its organs would have no functions. "Such a double has no proper functions because its history is not right. It is not a reproduction of anything, nor has it been produced by anything having proper functions" (Millikan 1989[c], 292). Although the human's heart has a function, its exact duplicate pumping away in the duplicate body would have no function, and so on for the eyes and all the other organs. Even if the duplicate came about in some lawful way that did not fit Millikan's definition and was not produced by something with proper functions, its component parts and processes would have no functions because they would lack the required history, nor would original artifacts that the duplicate human made, even if made with deliberation and skill. Instead of treating this as a consequence sufficiently counterintuitive to prompt theory revision, Millikan retains her definition by accepting the consequence. If one is prepared to accept enough, of course, any position can be maintained.

We constantly make judgments about what the functions of something are without knowledge of its history. An anatomy text talks about the function of the various organs, glands, muscles, and bones of the body. These functions were discovered not by researching their history but by examining their structure, their relation to other parts, and their effects, often by seeing what happens when the part is removed or rendered ineffectual. Knowledge of the functions of the human stomach had to wait until 1822 when Alexis St. Martin had his blown open by a musket blast, creating a permanent fistula convenient for William Beaumont's extended observation and experiment. Where history is needed to identify something, as it may be in identifying counterfeit money, documents, paintings, or pottery, that history must, on pain of circularity, be ascertained independently, not inferred from the item being investigated, then causally linked to the item in question. A historical procedure of this or any other kind is noticeably absent in determining functions. On the other hand, if Millikan's proper functions are quite distinct from the biological functions known and described for centuries, then appeal to the latter is inappropriate; and if proper functions overlap biological functions, the dimensions of the overlap should have been revealed.

Although the definition also covers artifacts, it accommodates only copies, not originals or one of a kind. "Artifacts that have been serving certain functions known to those who reproduce them and that are reproduced on this account (e.g., household screwdrivers) have these functions as direct proper functions" (Millikan 1984, 28). Eli Whitney's second cotton gin had a direct proper function, but his first did not, even though the second, let us suppose, was like the first. So that the first in a series should not be entirely bereft of function, Millikan gives it a derived proper function. Identifying the relevant direct proper function from which it is derived is left uncomfortably vague, but with a suggestion that it is something having evolutionary origin, presumably linked to survival. This leaves the function of the prototype, on which all the attention, thought, and planning were directed, with an obscure pedigree, while mere copies carry the certifying stamp of direct proper functions, from which all other functions come. In addition, many artifacts are not reproduced at all. A bird feeder built on the basis of the materials available, the kind of bird desired, and the need to keep out the rain, but not copied, is given only a function at second hand, one derived from we are not sure what. Moreover, much of what we produce has nothing to do with the agent's survival, as doorbells and tablecloths, making it dubious if the required direct proper function can do the job ordinarily ascribed to goals. Further, since there are two sources of an artifact's proper functions, they need not be the same. This Byzantine treatment of the function of artifacts, which most analysts regard as so uncomplicated as to give it paradigm status, gives Millikan's account an air of theory-driven unreality.

Because the history requires that the ancestors of m actually perform F, no way is provided for an artifact to have a function when the ancestors, the early versions, did not or could not work. One thinks, of course, of perpetual motion machines or elixirs of youth or spaceships built to travel faster than light. Neither they nor their ancestors worked, yet they certainly had their respective functions. These examples might suggest that such cases are rare and exotic and can, therefore, be safely ignored. On the contrary, it may be common for artifacts, especially machines, to be crafted to do a task that, from prototype to final version, they never do or never can do. We infer that such consistent failures are rare because we do not hear of them, but it may be instead that we do not hear of them because they were consistent failures.

The definition requires that the ancestors include both members with C correlated positively with F and members without C, suggesting Mill's Methods of Agreement and Difference. Suppose pressing a bar is m, an electrical connection between the bar and an electromagnetic latch on the pellet box is C, and producing food is F. Breaking the connection would be removing C. Learning may occur with the connection left intact and all trials ending with food, that is, without removing C. Although Millikan is at pains to give an account broad enough to include learned behavior, the account incorrectly excludes behavior learned with all trials reinforced.

Since history is forever, if functions are determined by history, functions are forever. New functions can be added, but old ones never die. That means that vestigial organs still have their original functions. It also means that the original functions of ancient artifacts that are still made, such as eating and cooking utensils, axes, knives, tents, blankets, bows, and arrows, are still present in currently manufactured versions, even if contemporary makers and users are unaware of them. It means that the crenels and merlons of public buildings still have the function of providing cover for archers and openings from which defenders can pour boiling oil. It also means that words still have their original meanings, along with all later meanings, which conflicts with the view of etymologists that meanings can be dated, hence, come and go.

The major difficulty of Millikan's analysis, however, is the one discussed earlier when examining analyses based on natural selection. In a summarizing statement, she says,

> A function F is a direct proper function of x if x exists having a character C because by having C it *can* perform F (1984, 26),

but immediately adds, "First interpret 'because by having C it *can* perform F' to mean 'because there were things that performed F in the past due to having C.'" When the passage is so rewritten, it becomes,

> A function F is a direct proper function of x if x exists having a character C because there were things that performed F in the past due to having C.

Rewritten with the intended interpretation explicit, it is surprising to find that it asserts that things are supposed to occur or ought to occur merely because they have occurred. She asks, "How could it ever be because it was *A* that caused *B* that *A* recurred as opposed to being, merely, because *B* occurred? How can a thing result from a prior *causing* as opposed to resulting from an effect produced via the causing?" (1984, 26). It can, she says, because, in part, As having caused Bs in the past sets up a positive correlation between As and Bs and it is that correlation that explains why As increased.

However, the correlation between As and Bs is not something above and beyond the fact that As caused Bs a certain number of times in the past. The most that can be inferred is that it might generate an increased probability that As will cause Bs in the future. Of course, if As caused Bs many times in the past, that gives some support for inferring that As will likely occur in the future simply because, for As to cause Bs, As have to exist. However, that has nothing to do with As causing Bs causing As to increase. The fact that lightning causes thunder sets up a high positive correlation between lightning and thunder; but that correlation does not cause more lightning.

Millikan describes her definition as a technical or theoretical definition. Under that umbrella, she gives her definition a status of studied ambiguity, ascribing features to it that make it both stipulative and descriptive, which she uses to block adverse comparisons with common or standard examples of functions (1984, 18; 1989[c], 289-93). However, for any account of functions to be of general interest and not a mere private excursion, it must connect with existing subject matter and existing terminology, in this case, subject matter and terminology that have been around for a very long time. Her motivation is to find the norms needed for a theory of meaning and reference in the norms present in biological functions, but that means that her theory, whatever else it is, must also be about pretheoretic functions, for that is where the norms she needs are found. In one of the statements on her definition, she says, "It may be read as a theoretical definition of function in the context 'The/a function of____ is ____' (the function of the heart is to pump blood), though *not* in the context '____ functions *as* a ____' (the rock functions as a paperweight)" (Millikan 1989[c], 291). If one takes this as her considered position, the various criticisms and counterexamples above are relevant, for they are all within the context of the modest restriction stated. Her occasional hints that she might mean her account to be based on analogy, as when she says that the purpose of the definitions is "to make as explicit as possible analogies among categories of things, which analogies had struck me as useful to reflect upon" (1984, 38), and her frequent use of mentalistic language in developing a naturalistic analysis of mental concepts, such as "the mechanisms within interpreter bees that are designed to translate observed bee dances into a direction of flight" (1984, 39) further complicate issues.

Dretske

Fred Dretske uses the concept of function as a fundamental part of his theory of representations on which to construct a naturalistic explanation of how reason influences behavior. A representational system is defined as a system having the function of indicating how things stand with respect to something else (Dretske 1992, 52). What something indicates about something else is the information it gives about it. There need be no receiver, nothing indicated to (Dretske 1992, 55).

Dretske divides representational systems into three kinds. In Type I, what does the representing, the symbol, and what it represents are entirely conventional. In his example, coins and popcorn represent basketball players. They do so in virtue of having the function of indicating the positions and movements of the players. This function comes about by being assigned or given by a person. Other examples are maps, diagrams, and road signs (1992, 53).

In the Type II representational system, there is a conventional component and a natural sign component. Natural signs take the place of symbols, indicate objectively, need no human intervention or reception, and cannot err. Quail tracks in the snow are natural signs. If the tracks are indistinguishable from pheasant tracks, they do not indicate that a quail made them. "As I am using these words, there can be no *mis*indication, only misrepresentation" (1992, 56). Representational systems must be able to err, and to do that there must be something the signs are supposed to indicate. This added element is supplied by functions and these functions are, as in Type I, assigned. Instruments and gauges and, apparently, control mechanisms are common Type II representational systems. In a thermostatically controlled heating system, the bimetal curvature indicates temperature (1992, 86-87). That fact that it does so is why the bimetal was selected to cause furnace ignition. The bimetal's being selected to cause furnace ignition because it indicates temperature gives it the function of indicating temperature. Whereas before it was selected, it merely indicated temperature, after it was selected, in addition, it now is supposed to indicate temperature (1992, 87-88). Another example is a fuel gauge. It represents what it is supposed to indicate, what it has the function of indicating. It indicates both the amount of fuel in the tank and the weight of the tank of gas, but it represents only the amount of fuel (1992, 59).

Type III representational systems are similar to Type II systems in that they are also based on natural signs, and are like both I and II in that the ability to represent and misrepresent comes from functions, but in a Type III system, the selection involved is natural selection. "The idea will be that during the normal development of an organism, certain internal structures *acquire* control over peripheral movements of the systems of which they are a part. . . . In the process of acquiring control over peripheral movements (in virtue of what they indicate), such structures acquire an indicator function and, hence, the capacity for misrepresenting how things stand" (Dretske 1992, 88). The Venus flytrap has hairs on each half-leaf. When an insect disturbs those hairs, the leaf snaps shut and digests the insect. The analysis is similar to that given for why the bimetal is supposed to indicate temperature.

> Once again, leaf movement (*M*) is caused by an internal state (*C*) that signals the occurrence of a particular kind of movement, the kind of movement that is normally produced by some digestible prey. And there is every reason to think that this internal trigger was selected for its job *because* of what it indicated, because it "told" the plant what it needed to know (i.e. *when* to close its leaves) in order to more effectively capture prey. (Dretske 1992, 90)

Other examples Dretske gives pertain to the mechanism's causing a plant to bud and form flowers in the spring and to the noctuid moth's hearing mechanism. "The noctuid moth's auditory system is obviously designed with its chief predator, the bat, in mind. . . . Why did the moth's nervous system develop that way? . . . The answer, obviously, is to enable moths to avoid bats" (1992, 91).

Although the Type I representational system seems uncontroversial, Types II and III raise questions. Thus, he says, in the case of the thermostat, that the bimetallic strip indicates temperature on its own, as a natural sign, but acquires the function of indicating temperature only when, because it indicates temperature, it is hooked up in such a way that it causes ignition. "The bimetallic strip is given a job to do, made part of an electrical switch for the furnace, *because* of what it indicates about room temperature. Since this is so, it thereby acquires the *function* of indicating what the temperature is" (1992, 87). However, the other parts of the thermostat, the mercury, the supporting structure, the wire and terminals, the cover, and so on, also have functions, and they were not selected because of what they indicated. Apparently they got their functions from simply being assigned them. Indeed, it seems strange to claim that coins and popcorn acquire functions simply by being assigned functions but that the bimetal does not. It seems far more likely that the bimetal gets its function the same way the popcorn and coins got theirs. However, if the bimetal gets its function simply from being assigned a function, Type II representation collapses into Type I, and whether or not the assigned function tracks natural signs becomes insignificant.

Linking natural signs to functions causes a further problem. Type II functions are based on natural signs, on what is, in fact, indicated. That means there can be no such thing as a function to indicate something that it does not actually indicate. Dretske says, "The indicator functions assigned an instrument are limited to what the instrument *can* indicate. . . . You can't assign a rectal thermometer the job of indicating the Dow-Jones Industrial Average. The height of the mercury doesn't depend on these economic conditions" (1992, 60). On the contrary, one certainly could, though it would be an outrageous error to do so. Error, however outrageous, does not yield impossibility. Artifacts that do not work but still retain their functions are all too common, as scissors that do not cut, and, among artifacts having indicator functions, clocks that run fast and speedometers that read slow. The point is neither new nor controversial, but is as essential to the concept of function as the ability to misrepresent is to the concept of representation. Indeed, the ability to misrepresent is just a special case of the ability to malfunction.

However, the greatest difficulty with Dretske's analysis of meaning based on biological functions is that he simply assumes the critical step on which everything else depends—that natural selection is able to create functions where before there were none. The description of the procedure is similar to that for Type II functions. C, which, in the case of Type III functions, is an internal state of the organism, indicates F, usually an external condition, as a natural sign. Over a long period of time, by natural selection, C "is recruited" to cause M (a response) because of C's indicating F, doing which gives C the function of indicating F. "Once *C* is recruited as a cause of *M*—and recruited as a cause of *M because of what it indicates about F*—C acquires, thereby, the function of indicating *F*" (1992, 84).

Indication is based on a dependency, usually causal, between the indicator and the indicated, such that if the indicated did not exist, the indicator would not.

In most cases the underlying relations are causal or lawful in character. There is, then, a lawful dependency between the indica*tor* and the indica*ted*, a dependency that we normally express by conditionals in the subjunctive mood: if Tommy didn't have the measles, he wouldn't have those red spots all over his face. (1992, 56)

Since indication is usually a causal relation, then when C indicates F, in most cases, it does so because C is caused by F and nothing else causes C, that is, F alone causes C or only F causes C. In Type III representation, C is recruited as a cause of M because it indicates F, thereby giving it the function of indicating F. Since we must not bring intentional concepts into a naturalistic account, saying that C is recruited as a cause of M must be understood as saying that C comes to cause M or, simply, that C causes M. Therefore, the account of Type III representation makes the unremarkable claim,

  C causes M because F alone causes C.

However, since the explanation of why C causes M must also be in naturalistic language, that explanation will have to be a causal explanation, producing the even more unremarkable claim,

  F's alone causing C causes C to cause M.

C cannot cause M unless C occurs and will cause M if it does occur, so whatever causes C can be said to cause C to cause M. Therefore, the above can be understood as saying,

  F alone causes C and C causes M.

There is nothing here that has anything to do with functions. Indeed, there is nothing here even to restrict the account to organisms. Suppose $rock_F$ rolls, hits, and dislodges $rock_C$ and, due to its size and position, it is the only thing that can do so, and that $rock_C$, now dislodged, rolls into and smashes $rock_M$. That would make it true that $rock_F$ alone causes $rock_C$ to be dislodged and $rock_C$'s being dislodged causes $rock_M$ to be smashed, or, in other language, that $rock_C$ causes $rock_M$ to be smashed because $rock_F$ alone causes $rock_C$ to be hit. Natural selection takes long periods of time, of course, but duration is not a part of the theory. This example appears to be a caricature of Dretske's account of Type III evolved functions, but that is because Dretske's account uses language that blurs the distinction between the mental and the physical. It is not the language of physics. Old habits of analyzing teleology with terms carrying intentionalistic overtones still flourish, even in the best circles, carrying with it the usual dangers of metaphor and circularity.

Papineau

David Papineau has also turned to biological functions for understanding language. He sees the fundamental problem to be that of explaining how something can represent something else and argues that beliefs, rather than neural states or causal patterns, can do so, provided the beliefs are linked to functions. "I want to argue that representation is best understood as a matter of the biological *functions* of our beliefs and desires" (Papineau 1987, 63).

> The reason the belief in question represents the presence of a *tree*, rather than a *tree-or-a-good-replica*, is that, although the belief is actually present more often in the latter, disjunctive, circumstance, its biological function is specifically to be there in the presence of *trees*, not tree replicas. (Papineau 1987, 63)

Like Millikan and Dretske, Papineau sees natural selection as generating these functions, and, also like them, extends natural selection to include learned beliefs. "Natural selection occurs within generations, by learning, as well as between generations, by genetic changes" (1987, 66). Among many random beliefs, an occasional one will lead to rewards and will become a fixed belief. "Only those few doxastic mutations which produce beliefs that are *usually* caused by circumstances in which the resulting actions are rewarding will get fixed" (1987, 66). Those circumstances become the normal circumstances, for example, the circumstance of a tree being present when the belief occurs that there is a tree. A false belief occurs when a belief is caused by abnormal circumstances. "Tree beliefs can be caused by tree replicas as well as by real trees. And in such 'abnormal' cases the relevant belief is false" (1987, 89). The belief is false because it is supposed to occur only when a tree is present, and that because its function is to be about a tree.

For natural selection to occur among learned beliefs, the beliefs must, he says, first occur randomly. He does not address the question how there can be beliefs before the belief-forming selection has done its work. If they already are beliefs, they must already be about something, in which case selection is not needed; if they are not already beliefs, selection cannot be among beliefs.

Like Millikan, Papineau bravely accepts the consequence that giving a natural selection foundation for beliefs to be about anything implies that a sudden replica of a human, even though having the same neural states, cannot have beliefs about anything. Also, like the others, he does not argue for a natural selection origin of functions, but merely assumes, though less confidently than do Millikan and Dretske, that most fundamental and critical step, that, though a new occurrence of a feature has no function, later occurrences somehow do.

Plantinga

Alvin Plantinga has recently erected an epistemology on the foundation of functions. In addition to the common functions, as that of the heart, there are also cognitive functions. "The purpose of the heart is to pump blood; that of our cognitive faculties (overall) is to supply us with reliable information: about our environment, about the past, about the thoughts and feeling of others, and so on" (Plantinga 1993, 14). Just as a damaged or malformed heart may not be able to perform its function well, a damaged cognitive faculty may not be able to perform its truth-acquiring function well. Furthermore, just as there must be an appropriate environment for the heart to function properly, there must also be an appropriate environment for the cognitive elements to function properly. Judging something to be red when it is not may be due either to malfunctioning cognitive equipment, as defective vision, or an abnormal environment, as illumination by red light.

The cognitive function, along with all the other functions and the conditions of an appropriate environment, are parts of a coherent encompassing plan Plantinga calls a design plan. "The design plan of an organism or artifact specifies how it works when it works properly: that is (for a large set of conditions), it specifies how the organism *should* work" (1993, 22). The cognitive faculties are a part of it. "What confers warrant is one's cognitive faculties working properly, or working according to the design plan *insofar as that segment of the design plan is aimed at producing true beliefs*" (1993, 16)." Reliability is brought in by requiring that the design plan is a good plan, specifically, "that the objective probability of a belief's being true, given that it is produced by cognitive faculties functioning in accord with the relevant module of the design plan, is high" (1993, 17).

Plantinga applies all this in a detailed examination of the issues regarding skepticism, memory, perception, a priori claims, other minds, testimony, Gettier examples, the Hume and Goodman problems of induction, probability, and rationality. He argues that the solutions lie in recognizing that humans have many proper functions, among which is the function of achieving truth and avoiding error. Does perception provide knowledge?

> When our perceptual faculties function properly, when they function in accordance with our design plan, we form *that* sort of belief in response to *that* way of being appeared to. Given an appropriate epistemic environment and given that the module of the design plan governing perception is successfully aimed at truth, such beliefs will have warrant; when held with sufficient firmness, they constitute knowledge. (Plantinga 1993, 99)

The versatility of this approach is surprising. However, although Plantinga appeals to functions, he does not analyze them. He considers his account a naturalistic theory, since functions are an entirely natural part of humans.

Chapter 8

Intentionality

Nearly all who have written on teleological language in science over the past several decades have approached the subject with the conviction that teleological language, if properly understood, does not imply or assume mind or design, and that the only problem remaining is how to work out the details. Nagel expresses this conviction clearly: "We shall therefore assume that teleological (or functional) statements in biology normally neither assert nor presuppose in the materials under discussion either manifest or latent purposes, aims, objectives, or goals" (Nagel 1961, 402). Canfield, in a later essay on teleology, agrees.

> There is thus a considerable natural inclination to regard talk of biological function as implying a conscious agent, endowed with the ability to form ideas of goals. Such a "teleological" interpretation of the biologist's language is no sooner formulated than rejected, for it ill matches the facts he reports on when he speaks of "function," "purpose," and "role." Organs have functions, but they are not conscious agents, and it has been a long time since the biologist per se could assume some artificer who designed animals and their organs. (Canfield 1990, 29-30)

However, since all the analyses studied fail in many and various ways and exhibit defects far beyond what tinkering can remedy, it ought not seem irrational or medieval at least to consider the familiar data of teleology without that presupposition.

Goal-Directed Behavior

Surely the place to start is to notice that the general teleological terms form a close-knit family. There seems to be no shift in meaning as one goes from "Hannibal's plan in crossing the Alps was to conquer Rome" to "Hannibal's purpose in crossing the Alps was to conquer Rome," "Hannibal's aim in crossing the Alps was to conquer Rome," "Hannibal's goal in crossing the Alps was to conquer Rome," and "Hannibal's reason for crossing the Alps was to conquer Rome." Although "end" is not used as frequently, there also seems to be no meaning shift in saying, "Hannibal's end in crossing the Alps was to conquer Rome." Finally, these are all expressible in "Hannibal crossed the Alps in order to conquer Rome." It is, therefore, unwarranted to select one of the terms, for example, "purpose," as Wimsatt does, or "goal" as several do, as being safe to use as a nonmentalistic term with which to analyze the rest. Such analyses appear to succeed, when they do, in part because they covertly rely on mentalistic concepts.

Although we say that Hannibal's plan, purpose, or aim was to conquer Rome, we do not mean that his plan, purpose, or aim was identical to his conquering Rome, for, although the former existed, the latter never occurred. One, therefore, must distinguish between plans, purposes, and aims and the situation or state of the world which they are about. Setting aside the question of the nature of minds, it does not seem extravagant to say that plans, purposes, and aims depend on minds. Since plans, purposes, and aims can also be called goals, these remarks apply to goals, as well. Thus, although we say that Hannibal's goal was to conquer Rome, we do not mean that his goal was identical to his conquering Rome, for a goal, like a plan, is a state of the subject about the object. The possessive in the expression "Hannibal's plan" or "Hannibal's goal" plays a dual role, for it not only identifies the plan or goal, but also indicates its dependent mode of existence. In this, it is unlike the expression "Hannibal's sword." His sword could exist by itself, but his plan could not. Conquering Rome could be a goal "for him," but it could not be a goal without his or another agent's having it.

However, since our language admits such sentences as "Hannibal's goal was to conquer Rome," or even, elliptically, "Hannibal's goal was Rome," it is easy to slip into the incorrect view that a goal is identical to a state of affairs or an event, such as conquering Rome, or even a physical entity, such as Rome. Nagel does this when he talks about plants being goal-directed systems. A goal-directed system must be a system whose behavior is directed to a goal, and that requires a goal. Since he is operating under the nonmentalistic assumption, he must mean by "goal" something not mind-dependent, that is, an objective state of affairs, a certain arrangement of the nonmental world, such as photosynthesis actually occurring, roots actually reaching moisture, or leaves actually facing the sun. This makes it necessary to refer to the mind-dependent interpretation as an "end in view" or "deliberate goal."[1] Nonetheless, as already noted, an objective state of affairs in the world is not a goal. The mere fact that plants regularly extend roots to moisture does not make their behavior goal-directed any more than the fact that rocks regularly roll downhill makes their behavior goal-directed. Nagel does not, of course, intend to imply that something, such as the plant, has the goal when he says plants are goal-directed, but that means only that he uses the term "goal" in the expression "goal-directed behavior" inappropriately and incorrectly. It is almost impossible for one to use "goal" consistently in the neutral way most analysts claim to use it. It is interesting to speculate how difficult it would have been for him to have used the term "plan" in this way and to have talked about plants engaging in "plan-directed behavior" while cautioning the reader not to interpret that as implying there was anything that had the plan. The same can be said about Wimsatt's use of "purpose" and the use of "goal" by Wright and others.

Ducasse, quoted earlier, in his familiar analysis of purpose, says,

> To be able properly to speak of an act (or event) as purposive, it is neither necessary nor sufficient that the act be such that unless it occurs some specified result will not occur. What is essential, on the other hand, is that the following elements be present, or be supposed, by the speaker, to be present:
> 1. *Belief* by the performer of the act in a law (of either type), *e.g.*, that If *X* occurs, *Y* occurs.
> 2. *Desire* by the performer that *Y* shall occur.
> 3. *Causation by that desire and that belief jointly*, of the performance of *X*. (Ducasse 1925, 153-54)

However, as discussed in examining Woodfield's purposive and artifact function TDs, we do not always desire to do what we intentionally do. Arthur Collins observes, "The universal availability of 'I wanted to . . .' should be evidence against the idea that it alludes to a prior state, since satisfaction of a desire is one among many alternative backgrounds for action" (Collins 1984, 366).

> We are wrong if we say that "in order to turn on the light" only explains in virtue of a tacit "I wanted to turn on the light." In the sense in which "I wanted to . . . " is always available, it is merely another way of

saying "In order to . . ." and not a further premise about prior states. (Collins 1984, 367)

Furthermore, if all our actions were really desired, we could not explain why we did a certain action by citing desire—something we, in fact, routinely do. We may explain our selection of a certain food or item of apparel or our viewing a work of art or a sunset simply by citing desire. If all actions performed were desired, it would be uninformative to say so.

In contrast, citing intentions does not explain actions. Michael Bratman observes that "intentions do not provide reasons that are to be weighed along with desire-belief reasons in favor of one considered alternative over another" (Bratman 1987, 34). Intending to do something settles the issue whether to do it and disposes one to consider the means, while merely desiring to do something accomplishes neither. Intention, because of its settled character, provides a foundation for further practical reasoning and decision making (Bratman 1987, 18-19). He sees intentions as conduct-controlling pro-attitudes in contrast to desires, which are merely potentially conduct-influencing pro-attitudes (1987, 22).

Finally, it is irrational to have incompatible intentions, such as intending to attend a concert and a lecture scheduled for the same time, but it is not irrational to desire to attend both.

Summarizing, intentions are not reducible to desires because we sometimes intentionally do what we do not desire to do, intentions do not provide reasons for actions while desires do, intentions settle issues, at least provisionally, but desires only influence them, and that only sometimes, and holding incompatible intentions is an error, whereas having incompatible desires is not. What is needed for an analysis of goal-directed behavior is one founded on intentions, not desires.

It is not enough, of course, to say that X does Y for the purpose Z simply because X does Y intentionally. The intention must be the right kind, namely, to produce Z. Further, X must do Y not merely intending that Z is produced, but must do Y intending that his doing Y will produce Z. The proposal, then, when the agent is the source of the intention, is that

> X does Y in order to do Z
>   if and only if
> X does Y intending thereby to produce Z.

Hints of the important role of intention can be found all through the literature. It is common among those writing on teleology to treat descriptions of intentional behavior as benign and uncontroversial, even to the point of giving it paradigm status, though, of course, restricting its application. Intentionality is the common teleological thread connecting goal-directed behavior and functions, explaining simply and clearly why a distinctive subdivision of our language, teleological language, spans both. Michael Levin, writing on how archeologists identify tools, writes, "But that human agency played a role in the creation of a physical object is insufficient for counting that object as part of material culture. First, the object must not have been created inadvertently—soot from a hearth is not material culture" (Levin 1976, 227).

The reason why intentionality is central to goal-directed behavior is that the behavior is mediated not by the realization of the goal but by the representation of the goal-state. Other accounts are unable to provide for the future orientation of teleological language while avoiding the implication of reverse causation, even though future orientation is widely regarded as a prominent, even distinguishing, feature of teleological language. Representation appears to be necessary to have future reference without reverse causation. Alan Montefiore observes,

> In teleological explanation we have a stimulus leading to the holding of a conception, which in turn only leads to the relevant behavior via its necessary forward reference to the goal state, G. . . . Here the conception is no mere accompaniment with no further causal contribution of its own to make in the production of G of which it is the conception; it is on the contrary an integral link in the teleological causal chain. (Montefiore 1971, 189)

It was noted earlier that much purposive behavior is performed without consciousness of that behavior, for example, habitual action, quick reaction, and actions that are the components of larger actions, making some of the standard criticisms against understanding teleology in terms of intentionality inappropriate. An intentionality analysis is often dismissed by assuming, quite incorrectly, that intentions must be conscious. Not so. Consciousness seems not to be analytically very significant, even in action theory, though intentionality certainly is.

Allowing for unconscious intentions is not an unusual position and should not be thought of as originating with psychoanalytic theory. Even the Ducasse analysis above does not specify consciousness. Typical is May Brodbeck's observation: "The difference between the action of raising my arm and the movement of my arm going up is that the former is done intentionally or for a purpose. The purpose need not be conscious; an action may be performed habitually, as reaching for a cigarette, or spontaneously, as jumping to dodge a car" (Brodbeck 1963, 320). Hempel, as noted earlier, makes the same point: "First: in many cases of so-called purposive action, there is no conscious deliberation, no rational calculation that leads the agent to his decision" (1962, 29). Many have remarked that beliefs and desires need not be conscious, as well. Woodfield's observation bears repeating:

> There is no need for an agent to be aware of the operation of his desire and belief when he acts purposively. He need not even know that he has them. . . . But it is well to remember that desires and beliefs need not be conscious. Since human beings can be in certain mental states without being aware that they are, it would be wrong to deny similar kinds of mental states to animals purely on the grounds that they are not aware of them. (1976, 171-72)

Myles Brand notes that routine actions, though intentional, are often not conscious (Brand 1986, 215). Although it is not necessary to say how consciousness fits into the scheme, it seems a reasonable conjecture that what we call the mental, including the intentional, is that which could occur consciously.

This analysis, therefore, is applicable only to goal-directed behavior in which the agent is the source of the intention. Such behavior might be said to exhibit internal teleology, a term already in use, though with various meanings. Internal teleological behavior, of course, requires that X be able to have the relevant intention and that the intention can influence the behavior. Most human goal-directed behavior, such as going to the store to get bread, falls into this category, as well as much goal-directed behavior of the higher animals. This category is similar to that described by Woodfield's purposive TDs. It does not include instinctive behavior or any behavior of lower organisms incapable of having the requisite intentions.

Most analysts simply ignore an analysis based on intentionality because it is not viewed as a serious alternative, but a few have bothered to marshal arguments against it.

Faber, as we saw earlier, argues against an intentionality analysis by saying that it is not needed; the job can be done by the right kind of machinery. "Can an organism's or a mechanism's goals be identified by reference solely to the structure and activities of the thing itself, apart from the intentions of its employer or designer? . . . I respond with qualified affirmatives to these questions" (Faber 1986, 76). It turned out, however, upon examination that reading goals from the structure of the mechanism, as a thermostat, resulted in all sorts of states being included as goals that no one wishes to include, as temperature stability in obscure and unimportant locations, stability in the dimensions of nearby metal fixtures, even maintaining a high level of heat radiated from the house, something desired not to occur.

One might, however, agree with Woodfield that one should not stop with intention, but analyze it as well.

> But since I am concerned with what it *is* to act purposively, as well as with the purely semantic problem of paraphrasing TDs, I ought to be able to provide a deeper analysis of the concept of intentional action. It should be possible to provide a general account of goal-directedness which is at the same time an account of acting intentionally. (Woodfield 1976, 174)[2]

One might, in fact, argue that to analyze goals and purposes in terms of intentions is merely to offer synonyms and that that is not analyzing at all. To regard it as trivially true, however, requires conceding the analysis to be least true, something currently vigorously denied. After all, the analyses studied were careful to avoid reference to intention, though they frequently and readily mentioned goals. Although there is currently considerable effort devoted to developing a nonmental account of intentionality, it has not yet been achieved; indeed, for several investigators, the preferred analysis is one in terms of biological functions. Considering the size of the current effort to understand intentionality and the nature of mind, it is unrealistic to expect the solution to drop out as a by-product of another study. It may well turn out that intentionality and teleology are so closely related that understanding one involves understanding the other or, conversely, that if one defies reduction, the other will, as well.

Alexander Rosenberg argues that descriptions of intentional states do not admit substitution of extensional identities, and so intentional states cannot be identified by nonintentional properties, ruling out psychophysical reduction: "Regularities employing the intentional description of mental states can never be theoretically linked up with neuroscience, or with non-intentional behavioral psychology. Therefore, they must remain isolated from the remainder of natural science" (1986, 73). Again, "intentional states are not related by causal or structural regularities to physical states, either behavior or physiology. . . . Molecular biology will never generate a science of the mental, under the intentional terms in which it is now described" (1986, 75). [3]

Finn Collin, in a paper addressing the lack of progress in developing a science of human action, sees the main obstacle as being one of not understanding the peculiar nature of intentional terms. Natural kind terms, such as "heat" and "light," can, as knowledge advances, be replaced by theoretical counterparts of greater integrative power, allowing a statement, for example, about the sunlight melting snow to be replaced with a statement about electromagnetic radiation and molecular motion. This can be done because light is identical to a certain kind of electromagnetic radiation and heat is identical to average kinetic molecular energy. In contrast, he says, terms such as "belief" and "desire" are not natural kind terms.

> They contain no blank for empirical science to fill out but already specify their extension in terms of the "essential nature" of the states in question, namely the intentional object of those states. For the generic identity of an intentional state is a function of the identity of its intentional object, the way that the generic identity of a chemical substance is a function of its atomic structure: The belief, desire etc. that p is identical with the belief, desire etc. that q if and only if p and q are analytically equivalent expressions. . . . Hence there is no such thing as a theoretical redescription of an intentional state which may replace the everyday specification of that state in explanations of action which are compatible with their everyday counterparts. (Collin 1987, 350)

Regardless how the study of mind and intentionality eventually goes, the intentionality analysis of goal-directed behavior fares quite well in the face of the many difficulties and counterexamples to which previous analyses proved vulnerable. Analyses that need the terminating event to identify goal-directed behavior are vulnerable to the missing goal problem. However, if goals are plans, aims, purposes, etc., and, therefore, states of the subject about the object, there is no problem about missing goals. In fact, the expression "missing goals" is misleading, for what is missing is not the goal but rather the achieving of it. Having a goal that is not achieved may be a practical problem, but it is no more a philosophical problem than is having a plan that is not carried out or having a wish that is not fulfilled.

Although analyses that linked goal-directed behavior to success in a probabilistic manner can accommodate occasional goal-failure but not extreme or necessary goal-failure, an intentionality analysis handles both equally well. Since goals are understood as intended results of some action rather than actual results, they are not limited by what is physically or logically possible. This analysis allows for the fact that one can have as a goal traveling faster than the speed of light or building a perpetual motion machine as easily as one can have as a goal going to the store to buy bread.

The multiple-goal problem, that the criteria used, such as plasticity or necessary condition, do not discriminate between two or more goals, is resolved because goals are determined by intentions, not behavior. If one is uncertain whether intentions can be attributed to cats, Scheffler's example can be modified to that of a person whose behavior of leaving for the woods with camping gear would be plastic both to fishing and to enjoying nature's solitude. Although the behavior would not determine that his goal is, say, the latter, his intention would.

The analysis provides for the forward orientation that seems so essential for teleology. The intention is about some later thing or situation. Yet, reverse causation is not a problem, for the intention precedes the action. Although the analysis is not committed to a causal relation, if the relation between intention and action were causal, the cause would be prior. In any case, the actual consequence of the action, such as buying the bread, is not claimed to cause the action that preceded it. It is not a consequence-etiology theory.

This analysis allows for the agent to have a goal and not act on it, what Woodfield calls disengaged goals. It also allows for goal change without behavioral change, something behaviorist theories cannot manage.

Teleological judgments are not restricted to expressions involving such words as "purpose," "goal," "aim," "in order to," and so on. To say that something pursues, chases, stalks, hunts, flees, escapes, waits for, tries, hides, threatens, courts, displays, shields, covers, conceals, protects, migrates, imitates, saves, stores, or searches is to use teleological language. Pursuit, for example, is following in order to overtake or capture, making "The dog pursued the rabbit" a teleological claim. Michael Simon observes, "To describe what the mouse is doing in terms of trial and error is to employ a teleological mode of discourse" (1971, 191). Teleological language is extremely widespread. It is difficult to describe behavior without it. As Michael Root notes, describing behavior by means of action words involves attributing intentions. Some words, like "courtship," obviously do, but, he adds, "verbs like 'invests,' 'forages,' 'harvests' also imply the bodily movements are intentional. The behavior of species distant from us or unrelated to us cannot be described using the terms of our ordinary language of behavior without imposing on them our psychology of intentions" (1989, 190).

As noted earlier, Wright points out that teleological judgments are often made quickly and with great accuracy and with no knowledge of the internal structure. He might have added that they are made by observers of low as well as high intelligence. It is, therefore, unrealistic for teleological statements to be analyzed in terms of complex internal structure, such as negative feedback. Negative feedback could be necessary for goal-directed behavior, but it cannot be a criterion for making correct teleological judgments. Although goal-directed behavior may involve feedback mechanisms, such mechanisms cannot be what teleological statements refer to or describe. Furthermore, if a person can be said to have made the teleological judgment that a dog is pursuing him even when he does not express that judgment verbally but only behaviorally by breaking into a run, it does not seem unreasonable to say that an animal that cannot express the judgment verbally makes a similar teleological judgment when like circumstances are followed by like behavior. Indeed, we even say such things as, "The rabbit saw the dog chasing him," thereby, rightly or wrongly, attributing a nonlinguistic teleological judgment to the rabbit. Since we are confident that rabbits know nothing about negative feedback, we certainly cannot be understood as attributing such knowledge to them. Similar remarks can be made about conceptually linking teleological judgments to evolutionary history. Attributing intentionality, however, to behavior seems a primitive move, as primitive as describing teleologically. Children do it with no instruction, as when told nursery stories and play-acting with toys. It is done in an instant, without conscious or deliberate inference.

This analysis of teleological descriptions of goal-directed behavior in which the agent is the source of the intention can be applied only to those nonhuman organisms that can be assumed to have the requisite mental capacities. Most people feel confident in assuming such abilities in the mammals generally, certainly the higher ones. Anscombe expresses what seems to be the general view: "Intention appears to be something that we can express, but which brutes . . . can *have*, though lacking any distinct expression of intention" (1957, 5). Brand outlines a plausible breakdown:

> We have common beliefs that persons act intentionally and that, as we descend in the order of life forms, the likelihood of intentional action decreases. Chimpanzees, it would seem, act intentionally, and perhaps also dogs and maybe—though doubtfully—white rats and pigeons. Our

intuitions are quite clear that sponges and slugs do not act intentionally. (1986, 215-16)

For an intentionalistic analysis to be correct, it is not necessary that one actually have accurate knowledge of such capacities. When one uses teleological language, saying, for example, that the rat pressed the bar in order to get the food and regards the rat as the sole agent of the action, one implies and, therefore, claims the requisite degree of mental life. If, in fact, rats do not have such mental life, the consequence is that that teleological statement is inappropriate or false. The consequence is not that there must, somehow, be an interpretation of teleological statements about agent goal-directed behavior that does not presuppose that degree of mental life. The behavior of lower organisms will be analyzed differently and will be understood in the same way as the behavior of organs and organic processes.

Functions

A fundamental question dividing analysts is whether teleological language exhibits a fundamental unity or not, whether there should be a single overall analysis, perhaps with subanalyses, or distinctly different analyses for the different kinds of teleology. A good example of the second approach is the much admired analysis of Robert Cummins. Though he allows a mentalistic analysis for the functions of artifacts, he offers a sharply contrasting analysis for the rest of functions:

> x functions as a ϕ in s (or: the function of x in s is to ϕ) relative to an analytical account A of s's capacity to ψ just in case x is capable of ϕ-ing in s and A appropriately and adequately accounts for s's capacity to ψ by, in part, appealing to the capacity of x to ϕ in s. (1975, 762)

To function as something is quite different from the function being that something and they should not be confounded. Robert Burch observes, "A wing may function for a crippled bird as a limb of ambulatory locomotion, but walking is not a function of wings" (1978, 47). Philosophical interest has always focused on functions, not on functioning as. Further, the analysis incorrectly requires that functions have capability, that "if the function of something in a system s is to pump, then it must be capable of pumping in s" (Cummins 1975, 757). Quite the contrary, the function of the heart is to circulate the blood in damaged hearts as well as healthy ones. Performance affects functioning as, but not functions—good reason for keeping them separate. In addition, his analysis does not accommodate the functions of nonhuman artifacts, such as spiderwebs, burrows, and nests, which have functions but are not components of systems.

Cummins disperses functions liberally. A forest is a system, s, and within that system, rain, x, has the capacity to water the ground, o, which watering appropriately and adequately accounts for the forest being able to grow, Y. Though the example meets the terms of the analysis, it is not the function of rain to water the forest. The analysis inserts functions throughout the realm of the physical sciences, giving cold fronts functions in weather systems, rivers functions in drainage systems, and electrons functions in atomic systems, where literal function talk has been conspicuously absent since the seventeenth century. It is one of the few analyses around that gives even heart attacks a function. Though utilizing the analysis in his own, Paul Griffiths, in respect to a liver infested with liver flukes, acknowledges (ruefully, one hopes), "This is the Cummins-function of the liver relative to the capacity to die of fluke infestation" (1993, 411). There is no attempt to relate this analysis to other areas of teleology, as if that were of no concern.

In contrast to such a piecemeal approach, Mark Bedau offers an analysis that, though it has its problems, does provide for the unity of teleological language. He sees the distinguishing feature of teleology to be value, that is, goodness or benefit. The principal formula, though others are considered, is

> A Bs in order to C
> *iff*
> A Bs because [A's Bing contributes to Cing and Cing is good]. (Bedau 1992[b], 790)

The beneficiary may be the agent of an action, the organism containing that which does the action (appropriate for organs), the species, or the user of the item (Bedau 1992[b], 792). The good may be any kind of benefit: survival and flourishing are goods for organisms and species, but the good includes also what is useful (1992[b], 791), even whatever is desired (1992[b], 797).

A good consequence theory is threatened by goal-directed actions not aimed at the good and artifacts made or used to produce harm. It seems strained to claim, as Bedau does, that even the masochist aims at the good (1992[b], 791). If, as he also claims, being useful is a good, even instruments of crime and torture must be viewed as having good consequences. To claim that what one desires is also a good (1992[b], 797), apparently implying that one cannot desire what is harmful, is either false or, because "good" must otherwise be interpreted so broadly as to exclude little, makes the good consequence claim innocuous. That all goal-directed actions must somehow have good consequences in any sense worth maintaining is implausible.

The parts and processes of an organism have indefinitely many beneficial consequences, most of which no one considers functions. The heart circulates the blood and the circulation of blood causally contributes to a host of beneficial consequences, as seeing, hearing, and running. Though they fit the analysis, they are not functions of the heart.

To provide for failure in goal-directed behavior and functions, Bedau lists as a variation of the above

> A Bs in order to C
> *iff*
> A Bs because A believes the following telic structure: [A's Bing is a means to Cing & Cing is good]. (1992[b], 796)

However, since beliefs may be false, this formulation is not a variant but a competitor to the original, for it allows that the result aimed at need not occur and, if it does, that it is not good. "What is taken to be a good consequence might either not be a consequence or not be good" (Bedau 1992[b], 796).

Although a good consequence analysis of teleology has its problems, it is a good example of an analysis that addresses the issue of the unity of teleological language and offers a plausible solution. It is significant that many of the same teleological expressions can be used both for goal-directed behavior and for functions, such as expressions involving the word "purpose." Thus, we can say both "He went to the store for the purpose of getting bread" and "The purpose of a potholder is to prevent burning one's hands." This breadth of usage applies not only to descriptions of goal-directed behavior of humans and to artifacts but extends also to descriptions of goal-directed behavior of animals, including lower animals, and to the functions of organs. It is not unusual to find statements like "The purpose of flocking is protection from predators" or "The purpose of the ossicles is to transfer vibrations of the tympanum to the cochlea." Such breadth of usage makes it seem forced and artificial to offer an analysis of only a part of teleological language, such as descriptions of functions only, or worse, of biological functions only.

That there is a close relation between goal-directed behavior and functions is not surprising in view of the fact that some goal-directed behavior is directed to the goal of designing, producing, altering, or using items having functions. Within the vast range of human goal-directed behavior is a large subclass of behavior directed to the goals of making tools, clothing, buildings, machines, etc. These are made for various purposes and various reasons. Some tools are made for the purpose of splitting wood, some for hammering nails, some for pruning twigs; some clothing is made for the purpose of keeping warm, some for keeping dry, some for display, and so on. The function of the item produced is the same as the goal of the behavior that produced it. Surely it is not a radical step to conclude that the function of the artifact comes from the purpose of the artificer. Surely that is the primary reason for the surprising breadth of teleological language, for the common terms and patterns found both in descriptions of goal-directed behavior and in descriptions of functions. That is also why we can say interchangeably "The function of the knife is to cut" and "The purpose of the knife is to cut."

Since a function does not arise independently but has its origin in the goal of the manufacturer, the function of an item or a feature is whatever the manufacturer intends that it accomplish. The function of the claws on a carpenter's hammer is to pull nails, not because the claws pull nails but because they were made to pull nails. Of course, as examples of rocks used as doorstops and paperweights remind us, it is not necessary that something be made or altered to have a function. The rock has its function simply because of how it is used. The claws on a hammer would not have the function of pulling nails if such use neither occurred nor were intended. Thus, a function falls within the context of goal-directed behavior because it exists as a consequence of goal-directed behavior in designing, making, altering, or using something.

Because things have functions in order to further plans, purposes, goals, or ends, things having functions are the intended or supposed means to an end. That is why, when asking for the function of something, one is likely to ask what it is "supposed" to do. Indeed, Wright, himself, no friend of an intentionalistic analysis, uses such language in identifying the functions of both a defective artifact and a defective organ, saying, "This is what it is supposed to do" (1976[b], 112). Things that have functions are not to be identified with the actual means to the end, for things can have functions that fail, but are the supposed or intended means to that end. A key is the intended means to the end of opening a lock.

These observations can be summarized as follows:

The (a) function of X is Y
  if and only if
W intends that X does Y.

It will often be that W also intends that Y contribute to some further end. Thus, we might say that the function of the ignition switch is to turn on the starting motor, thereby to start the engine. However, function statements are often only two-termed, as in saying that the function of a cup is to hold beverages or a function of the spleen is to store red blood cells.

Saying that X's doing Y is intended makes Y a contemplated end or goal, so this position claims that functions occur only in the context of goals. For W to intend that X does Y, W must have some role in respect to X, such as designing, making, altering, or using X.

Linking functions to goals is widely and emphatically rejected. Michael Simon, who takes a somewhat Kantian view of functions, explains the reason for such linkage.

> The fact that functional analysis depends on prior selection of the system that is supposed to be benefited shows that the notion of function is essentially an extrinsic one. Before we can attribute a function to something, we must have determined what ultimate goal it could be expected to serve. . . . To seek to know the function of something presupposes knowing what having a function would consist of, and this depends on having a preconception of what system that something is supposed to serve, if it is indeed functional. (M. Simon 1971, 83)

This explanation seems correct, except that the goals need not be beneficial and need not be ultimate. Woodfield, before he offered his own general analysis that tied lower life functions to benefit rather than purpose, stated, "A necessary condition of functionality is that the item being explained should serve some purpose for a thing other than itself" (1973, 39). In an essay on how to understand function attributions, Berent Enç and Fred Adams argue that the solution "lies in the acknowledgment of a broader taxonomy of properties (or behaviors) classified together by reference to a distal goal" (1992, 653). Kant's regulative view of functions links functions to purposes and goals, though giving them an obscure ontological status. Clark Zumbach notes that, according to Kant, we view biological wholes as determining the parts, as when a crayfish regenerates a lost claw, but he then adds that human knowledge requires that data conform to the second analogy, the causal principle, concluding, "But if we wish to make sense of the effect 'determining' its cause, we must conceive of it as being *intended*. For *an intention that this effect occur* can happen in time prior to the cause of the effect" (1981, 71).

Some analyses that avoid explicitly linking functions to goals do so anyway, using other terms. Elizabeth Prior holds that "the account of functions . . . is interest relative" and sees our identifying survival and reproduction as the biological interest of all life as due to our interest in our survival and reproduction (1985, 325-26). To illustrate the interest-relative nature of function, she imagines a world in which certain beings begin life in human form, a life that is wretched and exploited. However, those who contract cancer are metamorphized in the terminal stage into a higher and happier life form. In this world, the interest might not be survival. "And in such a world there would be corresponding changes in what counted as malfunction (and non-function)" (Prior 1985, 326). Her reference to an interest is what is here referred to as a goal.

The controversial issue is whether teleological language is significantly altered as one moves from human artifacts to the parts and processes of organisms and to the behavior of lower organisms. Wright is surely correct when he says, "Functional ascriptions of either sort have a profoundly similar ring" (1976[b], 106). If one is willing to let the chips fall where they may, the evidence, quite frankly, seems to be on the side that says the concept of function is not altered but remains a stable part of that large family of folk psychology concepts that includes plans, goals, reasons, wishes, desires, and so on. Wright continues, "Any analysis that can accommodate them both as functions in the same sense automatically has a point in its favor; any analysis that cannot has a difficulty it must explain away" (1976[b], 106).

Since function language undergoes no noticeable shift in meaning when applied to a wide range of subjects, it seems only sensible to try for a unified analysis for all function talk. The principal objection, of course, is that doing so means treating the functions of the parts and processes of organisms like the functions of artifacts, implying some kind of artificer. This would presumably place biological functions outside the boundaries of natural science, doing to teleological language in the life sciences what was done to teleological language in the physical sciences long ago. Should that be surprising? The alternatives are not attractive—limit the analysis to a subclass of teleological language, as many do, ignoring the close relation to the rest of function language, ram an analysis through, ignoring its objections, modifying one's linguistic intuition so that what were formerly defects now pass muster and the hitherto obvious becomes dubious, use concepts from elsewhere in the closed circle of mentalistic language to analyze the rest, or appeal to metaphor whenever theory falters and counterexamples loom—a metaphor of the gaps approach. In spite of the weight of opinion to the contrary, it should not be regarded as exceeding the limits of rational discourse or responsibility to suggest that the features of teleological language that prompted so much dispute in the physical sciences in the seventeenth century might still be present in the life sciences in the twentieth. If certain consequences of teleological language are, as a matter of brute fact, present, it is only being forthright and realistic rather than medieval or obscurantist to acknowledge them. It is noteworthy that the only weighty and sustainable objection to a unified analysis of functions that has surfaced is the metaphysical one. No other analysis handles the traditional problems and requirements of function theory so effectively and smoothly.

As indicated, the intentionalistic analysis of functions offered earlier is presented as applying to all functions, including the parts and processes of organisms, especially organs. There remains, however, the problem of how to handle teleological language describing the goal-directed behavior of organisms believed not to have the requisite mental life to be the source of the needed intentionality. This includes all instinctive behavior and all the behavior of lower animals and plants: birds building complex nests, spiders spinning intricate webs, butterflies migrating to Mexico, flowers seeking the sun, and roots reaching for moisture. The general pattern found in the analysis of functions is extended to include such behavior, with the intention having its source outside the behaving agent. The modern notion of programming, though anachronistic, seems not inappropriate. It is like the analysis of functions except that what is intended is a goal-directed action. The pattern is:

> O does Y in order to Z
>   if and only if
> W intends that O does Y thereby producing Z.

"O" stands for an organism, plant or animal. The behavior is still intentional, but the source of the intention is outside the subject doing the behaving.

On this view, teleology divides naturally not into goal-directed behavior and functions, as many have read it, but into internal and external teleology. Functions fall entirely within external teleology, and much, but not all, goal-directed behavior.

Few have bothered to argue against an external intentionalistic analysis, presumably because it seemed limited to human artifacts and hence of little concern in science. Bigelow and Pargetter claim that some artifacts have functions with no representational or intentional component because they were created with no function in mind.

> Many artifacts evolve by a process very like natural selection. Variations often occur by chance and result in improved performance. . . . They may have been produced with an over-all function in mind (say, hitting nails); but the toolmaker may not have in mind any functions for the components and features of the tool, which contribute to the over-all function. For instance, the toolmakers may copy a shape that has the function of giving balance to the tool—but they need not foresee, or plan, or represent any such function. They know only that tools like this work well at banging nails, or sawing wood, or whatever the over-all function might be. Consequently, even with artifacts, structures can serve specific functions even though there exists no prior representation of that function. (1987, 185-86)

Suppose a tool is made for hammering and that that function is intended. As the authors see it, the hammer might have some feature that, unknown and unplanned, gives it good balance, perhaps a wooden rather than a steel handle, thereby giving it a feature having a function that is unintended. Their assessment, however, is incorrect. If no one, now or earlier, installed a wooden handle in order to give the hammer balance, used a wooden handle because it gave the hammer balance, contemplated so using it, provided it for another to use, etc., if the balance were just there, ignored and useless like a nick or a scratch, it would not be a function of the wooden handle to give the hammer balance.

Griffiths offers a similar argument:

> Many features of artifacts make no intended contribution and yet have proper functions. . . . Various shapes of a tool are tried out. One is more effective than the others because, for example, it is better balanced. This shape is copied more than the others and eventually becomes the norm in that culture. The function of the shape is to balance the tool. (1993, 418-19)

Far from showing the absence of intention, his example reveals its presence. Trying tools of various shapes, noticing that one is better balanced, and copying it for that reason is highly intentional behavior.

Faber also argues against an intentionalistic analysis, claiming that intention does not pick out functions. As discussed in detail earlier, he offers as evidence examples in which items, such as voltage regulators and thermostats, are produced accidentally, arguing that what the item does determines its function (1984, 87; 1986, 97-98). These examples were criticized on the grounds that Faber incorrectly assumed we really could imagine such complex machines constructed accidentally and also on the grounds that his examples had components exhibiting extensive design and so were far from being intention-free. In addition, Faber allows context and use to determine function, yet intention is a common part of context, and use is clearly an intentional concept (1984, 102-103). Finally, his report of how we would judge cases was incorrect, for we actually determine function by looking for intended use, not, as he alleges, by "reference solely to the structure and activities of the thing itself, apart from the intentions of its employer or designer" (1986, 76).

Christopher Boorse rejects grounding functions on intentions as well. Wright's broken windshield washer button, he says, has a function only in a Pickwickian sense, not in the normal sense, because it fails to work (Boorse 1976, 72-73). The issue is whether an item clearly designed to do Y has a function Y even though it fails to do Y. Wright is, of course, correct—the washer button still has its usual function. "Function" is not a success word, in contrast with "knowledge," which is. Boorse's view would remove the distinctive element of teleological language, its "supposed to" character. If a function were simply an effect, there would have been no need for teleological language.

It is dismaying how the claim that for X to have a function Y, X must succeed in producing Y comes up again and again in the face of overwhelming and conclusive objections to it. The labels on radio controls state the functions of those controls regardless of whether those controls work or not. If the repairman were asked to identify the function of a certain knob, he would reply that its function was, say, to control volume, and would do so without checking to see whether the knob actually did control volume. Schematics list the functions of various components. They could not do this if functions vanished when the component did not work. Since schematics are prepared for repair shops, it is expected that they will be consulted regarding radios with parts that do not work. The repairman does not correct his schematic, even mentally, subconsciously, or by implication. The schematic has no disclaimer notifying the reader that the function labels apply only when the component is in working order, nor does the technician read that in. The schematic is useful to the repairman as is, functions baldly stated, because it enables him to understand how the circuit is designed to work. The function of this control is to regulate volume both on this radio that works and on this one brought in for repair with a volume control that does not control volume. When pressed on what the function of something is, we frequently fall back on "It is supposed to" talk, where what follows the "supposed to" refers to the intended result, not the intended function, implying that when one has stated the intended result, one has stated the function.

If a wall switch is miswired and really rings the doorbell when it was supposed to turn on the hall lights, we do not say that its function is to ring the doorbell even though that is its effect. To suppose otherwise would make the concept of malfunction useless. A broken switch does not turn on the light and neither does a doorknob; but only one malfunctions when it does not turn on the light. Malfunctioning is not lacking a function, but lacking the result intended. The very concept of malfunction is evidence for an intentional analysis of functions. Wherever function talk is applicable, malfunction talk must be as well, and malfunction talk transparently points to what the item is supposed to do, to what it is intended to do.

In arguing the case that functions require success, Boorse describes an air filter designed to remove a certain pollutant from the atmosphere. Many years later, we are to suppose, this pollutant is no longer present in the atmosphere. Does the filter still have a function? "It seems," he says, "more natural to hold that although the filter used to have a function, it is currently nonfunctional" (1976, 74). However, "nonfunctional" means, in this case, not lacking a function, but not working, not performing its function. A defective switch may be described as nonfunctional, but that means that it does not work, not that it has no function. The terms making up the family of function language bear no simple relation to each other. Having a function is not the same as is being functional.

Adams, in a passage we looked at earlier in a different context, criticizes the intentional analysis of functions along similar lines.

> It seems one's intentions (i.e., the fact that one *selects* x for y) are neither necessary nor sufficient for x's having the function y. If I build a system S and include x, intending for x to do y (selecting x for y), that

is not sufficient for x's doing y. Unbeknownst to me x may not be able to do y. If I build windshield wipers out of cardboard, it is not sufficient that I selected them for their ability to clean windshields, that they have that function. If they cannot enter into the means-ends nexus in virtue of which my car outputs O (cleans the windshield in the driving rain), then that is not their function. (1979, 512-13)

As was the case with Boorse's example of the broken windshield washer button, Adams' intuitions are just wrong. The example is a little extreme since it requires us to imagine an error nobody would make, and this may give us a distorted sense of where the problem lies. A more believable example would be wipers made of rubber too hard to clean the glass or of material that smudged the glass. In either case, we would say that they did not work as they were supposed to, indicating the function as intended result.

Canfield offers a similar argument.

The tie in the original to intention and purpose is not as close as one might imagine. . . . In the case of the functions of artifacts, that something was designed to do so and so is neither necessary nor sufficient for its being the case that its function is to do such and such. For instance, something may be designed so poorly it is useless for doing what its designer had in mind, but a people may put it to use doing some other job for which it turns out to be well suited. (1990, 48-49)

On the contrary, as already argued at length, functions need not succeed to be functions. That is not peripheral to being functions but a critical and central feature. Uses involve means and ends, and using something requires intentionality quite as much as designing does. Uses do not occur by themselves, like colors and shapes, but require purposes and goals. People and animals use rocks, but rocks do not use anything. Canfield continues, "It is true it is *used*—that is the concept of an artifact's having a function—but its use does not entail planning or an even explicit thought of the use it is to be put to" (1990, 49). No doubt many times a use comes about without explicit thought or planning, as may occur when using a wrench to pound something or a screwdriver to pry things, though conscious origin is probably the rule. In any case, the difficulty with this argument to sever the linking of functions to intention is that it requires intention to be a conscious state, which we have seen is a recurring theme in arguments against intention. As already pointed out, however, intention need not be conscious.

Boorse relates that yeast was used in alcoholic fermentation long before it was understood how it worked. Only in recent times was it discovered that its function is to produce enzymes that catalyze the conversion of sugar to carbon dioxide and alcohol. Since the function was not known, he argues, it could not have been intended. Therefore, during that long period, there was a function without intention (Boorse 1976, 73).[4]

The example is interesting. Yeast used for making beer and bread has been found in Egyptian tombs built around 2000 B.C., but it was not known that the yeast produced alcohol by furnishing enzymes that acted on sugar until the work of Caignard de la Tour in 1857. The yeast, on this view, had that function for nearly four thousand years with no one knowing it. The example is supposed to show that the function of yeast is what it actually contributes to the fermentation, not what it was intended to contribute.

Catalyzing the conversion of sugar to carbon dioxide and alcohol was a component effect within the larger function of yeast to produce to fermentation. Where there is interest in this component effect and where it is also intended, it does not seem wrong to call it also a function. However, an effect does not have to be known to be intended. In 1795 the British navy ordered daily rations of lemon juice (lime juice came later) for its sailors to prevent scurvy. It did so by supplying ascorbic acid, but that was not known until Szent-Gyorgyi isolated ascorbic acid in 1932. Although the British navy did not know by what means lemon juice prevented scurvy, they surely believed it did so by some means or other, by some mediating effect, and that effect was, therefore, intended. In 1847, Ignaz Semmelweis had medical personnel in the obstetric clinic wash their hands in chlorinated lime before examining patients, and the death rate from puerperal fever dropped from 12 percent to 1 percent. Although he did not know by what means washing in chlorinated lime prevented the spread of the fatal disease, he surely believed it was by some means, and that means was intended. We take aspirin to relieve pain, but until very recently, no one knew how it worked. Yet, we intended, whatever the mediate effect was, that it occur. This is similar to the still current practice of injecting gold in managing rheumatoid arthritis. Early fermenters surely believed that yeast produced alcohol by some means or other, and that unknown mediating effect was intended. It should be noted that all these cases involve intentional actions.[5,6]

Intermediate causal links may or may not be functions, depending on interest. In these examples, the mediating effects were, even when not yet identified, fairly significant. People have wondered what it was about yeast or lemon juice or chlorinated lime or aspirin or gold that enabled them to have their desired effects, and so it does not seem unreasonable to call these mediating effects functions. However, causal links may be subdivided beyond the point of interest, as is talking about moving blood through the first bend of the brachiocephalic artery. Prior was right. Functions seem clearly to be interest relative.

Regardless whether past beliefs were correct or incorrect, we still attribute functions according to the beliefs of that day. Purges were long used in the belief that they improved health by restoring balance to the four humors by reducing the melancholic humor. Historical accounts describe the purge by saying its function was to reduce the melancholic humor, even though it did not do so, thereby linking the function to the intended effect, not the actual effect.

Achinstein also argues against an intentionalistic analysis. "The ends with which functions are associated can be *pure*, i.e., ends which no one or thing actually desires or intends" (1977, 360). He imagines a sewing machine with a button designed to activate an exploding mechanism. Neither the designer nor the user nor anyone else ever intended that the button be used. The function of the button is to activate the exploding mechanism, but there is no intention to activate that mechanism; hence, he says, there is function without intention.

To design the button to detonate the explosive is to make the button with the intention that it be able to detonate the explosive. It is quite possible to do that while not intending that the button be used to detonate the explosive. Bringing about the capability to do something is distinct from bringing about the doing of it, and either can be an intended objective. An army instructor might modify a sewing machine to use as a model of a booby trap for an antidemolition training class, and, to heighten realism, actually wire it to work, yet never intend it to be used. He did, however, intend that pressing the button would detonate the explosive, and that is sufficient to give the means intentional status, for it identifies its intended effect.

There are means to conditional ends as well to unconditional ends. Fire escapes are placed on buildings to provide a means to the conditional end of escape in case of a fire. Most buildings never burn, so for most buildings, the conditional end never becomes an actual end, since no one ever has the end of escaping the fire in those buildings. A fire escape, nevertheless, throughout a building's lifetime, exists as a means to this conditional end. Sandra Mitchell writes:

> The function of the foam rubber in the removable seat cushions on an airplane is to allow the passenger to float if the plane should go down over a large body of water. The floatation function of the foam in the seat explains why the foam is there even if the unfortunate (and presumably rare) event of crashing over water never occurs, so that the seat is never used in that way. What is crucial in the case of an artifact designed and created by intentional human action is the *belief* that the result—i.e., floating—would occur in the relevant situation, rather than that it actually occur. (1989, 218)

No one intends that the plane goes into the water, so no one intends simply that the passenger floats on the cushion, but manufacturers do intend that if the plane did go into the water, the passenger would float on the cushion. So also with Achinstein's example, no one intends that the button is pressed, so no one intends that the machine explode, but the builder did intend that if the button were pressed, the machine would explode.

A common theme running through several otherwise divergent analyses, including those of Canfield, Lehman, Wimsatt, Wright, and Woodfield, is that biological functions can be grounded on natural selection. As already discussed in detail, there are many difficulties with this proposal, but the fundamental one seems to be that natural selection cannot generate norms; it cannot provide the essential characteristic that distinguishes teleological language, the "supposed to" character of functions and goals.

There is indirect evidence in favor of an intentionalistic analysis of functions. One such piece of evidence is that function talk is intensional. Few have noticed the extent of intensional language in biology. Rosenberg, talking about a wide range of teleological language in biology in the course of arguing for a point of view very different from the one endorsed here, explains why intensional language is significant.

> Cognitive or intentional states have a logical property which distinguishes them from all other states. . . . It is a property which hinges on their representative character, on the fact that they "contain" or are "directed" to propositions about the way things are or could be. The property they have is this: when we change the descriptions of the states of affairs they "contain" in ways that seem innocuous, we turn true attributions of intentional states into false ones. Intentional states are intensional. By making innocent substitutions within intentional

descriptions of terms that refer to the very same objects we can turn truths into falsehoods. But this is something we simply cannot do to any expressions that report *non-intentional* relations, such as those characteristic of physics, chemistry or biology. (1986, 70)

These remarks are also applicable to function talk.

Eric Kraemer also observes that substitution of identities does not preserve truth-value in biological function talk and that failure of substitution of identities is a mark of intentional descriptions. Even though, for example, Smith believes Jones is prudent and Jones is identical to the junior senator, it does not follow that Jones thinks the junior senator is prudent. It is, he adds, not so well known that the same holds for functions.

(Q) The function of heart beating in mammals is to promote the circulation of the blood.

(R) The circulation of the blood is the activity most commonly used as a philosophical example of a biological activity.

(S) The function of heart beating in mammals is to promote the activity [most] commonly used as a philosophical example of biological activity.

From (Q) and (R) it seems that we cannot deduce (S). (Kraemer 1979, 151)

He then goes on to show the contrast with causal language.

(T) Heart beating causes heart noises.

(U) Heart noises are my favorite noises.

(V) Heart beating causes my favorite noises.

In this case it is clear that the inference from (T) and (U) to (V) certainly goes through. (1979, 151-52)[7]

The point can also be made in the case of artifacts. Although the function of auto engines is to convert fuel into mechanical energy and converting fuel into mechanical energy is the major source of air pollution, it is not the function of auto engines to be the major source of air pollution. On the other hand, because auto engines cause the conversion of fuel into mechanical energy and converting fuel into mechanical energy is the major source of air pollution, it does follow that auto engines cause the greatest source of air pollution. From consideration of examples similar to this, Kraemer concludes:

What I hope to have suggested here, then, is that there are certain
properties which are such that: (1) propositional-attitude sentences and
function-attribution sentences share them, (2) causal-attributions lack
them, and (3) these properties intuitively are expressed in the phrase
"directed on an object or state-of-affairs which need not exist." (1979,
151-52)

Achinstein, however, argues that singular causal statements are not
extensional. His examples rely on the fact that emphasizing a part of the
statement can change the meaning of the statement. Thus, "Socrates'
drinking hemlock at dusk caused his death" is, he says, true, while
"Socrates' drinking hemlock *at dusk* caused his death" is false because
the change in emphasis produces a change in meaning (Achinstein 1975,
3; 1979, 369). "The reason is that the emphasized words become selected
or captured by the word 'caused', indicating that a particular aspect is
causally operative" (1979, 371). However, to say something is causally
operative is to say that it is a cause or a part of a cause. Thus, to say that
the second sentence claims that the time of drinking the hemlock is
causally operative is to say that the time of drinking is a cause or a part
of a cause of death, something not claimed in the first sentence. That
means that the first sentence, since it mentions the time of drinking but
does not claim it to be a cause or part of a cause of death, talks about
more than what caused death. However, the claim that causal language is
extensional is a claim about causal language only; it not a claim about
sentences that talk about other matters as well. Of Achinstein's two
sentences, only "Socrates' drinking hemlock *at dusk* caused his death" is
a purely causal claim. Informal language, which is the language we speak,
is, well, informal, and, causal statements often occur in which extraneous
information is included, material not intended to be taken as part of the
cause.

The referential opacity of function claims over against the transparency
of causal claims lends support to the position that functions are a kind of
intentional entity belonging to the same family as desires and beliefs and,
incidentally, that it is unlikely that function language can be completely
reduced to causal language.

A feature of functions conspicuously ignored in the literature is that
functions have a certain breadth, a certain rough texture to them. We say
that the function of the eyes is to see, not to see food, that the function
of the heart is to circulate the blood, not to circulate the blood to the
duodenum, and that the function of the skull is to protect the brain, not
to protect the brain from falling roof tiles. A necessary condition or

requirement analysis or one based on survival or benefit does not make such distinctions. The eyes are necessary both for seeing and for seeing food, and both seeing and seeing food are beneficial and improve survival rates. The skull is necessary both for protecting the brain and for protecting the brain from falling roof tiles, and both are beneficial and improve survival rates. We notice, however, that we say the function of a pocketknife is simply to cut and do not say it is to cut rose stems, and that the function of a compass is to determine directions, not to determine north by northwest. The reason why the function of a pocketknife is simply to cut and not to cut some specific object and why the function of a compass is to determine direction and not a specific direction surely is because we imagine the designer or manufacturer had only general uses in mind. The fact that biological functions have this same coarse-grained property suggests that, disclaimers notwithstanding, they are also viewed as artifacts.

There is general agreement that viewing organisms as having purposes and as being designed has heuristic value. Even Hempel, who finds little useful in function explanations, says,

> One of the reasons for the perseverance of teleological considerations in biology probably lies in the fruitfulness of the teleological approach as a heuristic device: Biological research which was psychologically motivated by a teleological orientation, by an interest in purposes in nature, has frequently led to important results. (1948, 145)

Ernst Mayr, noting that teleological language uses the terms "purpose" and "goal" (1974, 91) and that "teleological means end-directed" (1974, 105), observes, "The heuristic value of teleological *Fragestellung* makes it a powerful tool in biological analysis, from the study of the structural configuration of macromolecules up to the study of cooperative behavior in social systems" (1974, 114). David Resnik describes in detail how functional concepts aided the work of Harvey on circulation, Watson and Crick on the model of DNA, and several in the development of the theory of the cell (1995, 123-132). Yet, scarcely anyone asks why this is so. One of the few who senses something amiss is Engels. She comments on the fact that analytic philosophy, including that of Hempel and Stegmüller, repeatedly points to the heuristic role of teleological language in research but at the same time restricts that role and affirms the translatability of teleological language into nonteleological language (1982, 198). Setting aside the many unresolved issues of inductive inference and confirmation theory, there is wide agreement among scientists, if not among philosophers, that if a hypothesis together with a description of relevant initial conditions imply that certain observable events will occur, events that would not have been expected were the hypothesis false, and those events do occur, they provide some degree of evidence for the hypothesis. It is unclear why this line of reasoning should be inapplicable to the heuristic success of the hypothesis of design when it is used so widely elsewhere. Surely there is something disingenuous with saying that parts of nature appear as if they were designed but really are not and, at the same time, holding that viewing them as designed aids research. Mere appearance, which even the Cartesian skeptic acknowledges, should have no value in guiding research. There is a marked absence of comment in the teleological literature on what appears to be a significant inconsistency.

It is interesting to look at the language used to describe something as not being a function. Wright says, "If the dimmerswitch makes a particularly good foot scratcher, that is only an accident; that is not why it is there and hence not its function," adding, "So 'accidental' is an insightful way of characterizing what it is about these things which prevents them from being the functions of those artifacts" (1976[b], 79). Adams writes of the problem of "fortuitous effects" that plagues some analyses of functions (1979, 499-500) and of those things that "just happen." "If a rock just happened to be thrown on Smith's desk and just happened to hold paper on his desk when a gust of wind blew in the office, we would still not ascribe to the rock the function of holding papers in place" (1979, 509). Lehman imagines a person shot in the brain: "By some quirk the bullet does not kill the person but instead relieves pressure that had been building up in that person's brain" (1965[a], 14). We contrast functions with side effects. We say that relieving pain is a function of aspirin but describe stomach irritation as a side effect, and we contrast the function of a machine with its by-products. Terms such as "accidental," "fortuitous," "just happens," "by some quirk," "side effect," and "by-product" in such contexts do not refer to what is random or uncaused, but rather to what is not planned, not intentional, and are so used also in ethical and legal contexts. Plantinga, in reference to a refrigerator that emits a loud squawk when a screwdriver touches a certain wire, writes, "This is an *accidental* or *unintended by-product* of the design. It is not accidental in the sense that it happens just by chance, or isn't *caused* to happen. . . . It is accidental, rather, from the point of view of the intentions of the designer" (1993, 24).[8]

It certainly must mean something that the terms used to distinguish effects of artifacts that are not functions from effects that are functions are the same terms used to distinguish the effects of organs that are not functions from effects that are. Regarding organs such as the heart, Woodfield writes, "Why is it that only some of an item's activities are functions, and the others accidental?" (1976, 108) and Ruse, "As things stand, heart sounds just seem to be a by-product of a beating heart" (1973[a], 188). Wright, regarding a stomach-intestine adhesion at the site of an ulcer, preventing discharge into the abdominal cavity, notes, "But no one would say that was the *function* of the adhesion; it *has* no function: it is only there by accident" (1976[b], 99). Regarding biological functions, Millikan says, "But the production of an accidental side effect, no matter how regular, is not one of a system's functions; that goes by definition" (1989[b], 283).

As noted earlier, functional statements are normally used to explain why an item or property is present. Thus, saying that the function of the ossicles is to transmit vibrations from the tympanum to the cochlea can be used to explain why the ear has ossicles. A function statement, however, could not have such explanatory power if it merely described the structure of the ear. A description of the shape, location, and substance of a part and its relation to other parts does not explain why that part is there. Saying that houses usually have rectangular apertures in the walls in which are set plates of glass does not explain why the apertures are there or why they are covered with glass. This, by the way, is also an argument against a negative feedback analysis, for it is also true that a description of arrangement of components of a negative feedback system does not explain why the system is there. Neither would a function statement have the explanatory power it has if, as Nagel and C. Taylor propose, it asserted a necessary condition or a requirement condition. Even if the heart were necessary for circulation, that would not explain the existence of the heart, just as the fact that rainfall is necessary for forests does not explain why it rains. On the other hand, if functions were intended effects and the subject having the intention played a significant role in the production of the item, that would account for why the presence of the item is explained by giving its function. If functions were intended effects and the manufacturer of claw hammers intended that the claws be used to pull nails, that would account for why the presence of the claws on hammers is explained by saying their function is to pull nails.

Some say that teleological descriptions of goal-directed behavior of lower organisms and biological functions are metaphorical. Although this would remove a host of problems, it comes at high cost. For one thing, teleological language is difficult to avoid, while metaphorical language is not. Rosenberg's comments about the extension of teleological language to talk about the macromolecules suggest its boundaries are expanding, not contracting: "The first and most important of these is its *extreme naturalness*: the behavior of these simple substances simply cries out for description in intentional terms. A second feature of this usage is that it is not only natural, it is inevitable and unavoidable" (1986, 66). Kant regards understanding organic life teleologically as unavoidable. Teleological language is certainly not dying out. However, if one holds that teleological language is necessary for describing biological phenomena, one cannot also hold that it is metaphorical. Metaphor may be useful and inviting, but hardly inevitable and unavoidable.

There is, however, a stronger argument against regarding the typical use of teleological language as metaphorical. As just noted, ascribing goal-directed behavior to lower organisms or ascribing functions to organs and processes is usually done to explain why the behavior or organ or process is there. A study is made of flocking behavior of certain birds when feeding and concludes that the flocking occurs for protection. If that conclusion is meant metaphorically, its explanatory claims vanish. It might be somehow illuminating to compare gray, cloudy weather to a somber mood, but that does not explain anything about the weather and does not show up in meteorological reports. Of course, as Wright argued, some metaphors have become standard usage and lose their metaphorical status. When that occurs, however, there is always an established and defined meaning of the formerly metaphorical expression. If "pursue" or "hide" or "flee" have undergone such a transformation so that there is one meaning when humans perform the actions and another, formerly metaphorical, meaning when lower organisms perform them, dictionaries would reflect that fact. They do not. "The man chased the bee" and "The bee chased the man" do not use different meanings of the verb.

Part of the indirect evidence for an intentionalistic analysis of functions is that it provides satisfactory responses in the many areas where other analyses have trouble. Some are unable to avoid the implication of reverse causation. As noted earlier, Wright, Woodfield, and Neander try to do it by restricting functional talk to types. This, as we saw, does not help, because abstraction does not affect causal order; it neither removes the causal ordering of the tokens nor reverses it. Furthermore, we do use function language about individuals as well as types. An analysis appealing not to consequences but to intended consequences does not imply reverse causation and accommodates function predicates applied both to types and to individuals.

A common problem facing an analysis of function is whether it can reject side effects. Ignoring the fact that heart sounds have diagnostic value and may affect the health of infants, heart sounds serve as the most commonly used example of function side effects. As noted earlier, since it is physically impossible in a normal organism for the heart to cause circulation without making heart sounds, this is a counterexample to construing functions in terms of necessary or requirement conditions. It is also a counterexample to an analysis based on effects that increase species survival. However, on the present analysis, heart sounds are rejected as being a function because it is believed, rightly or wrongly, that they are not an intended effect; indeed, that seems to be precisely what is meant in calling heart sounds a side effect. Of course, there can be error in judging an agent's intentions, but we do, as a matter of fact, routinely infer intentions from artifacts. If there were an error, that would mean only that there would be a corresponding error in the function attribution, and not, as some conclude, that such inferences ought not be made or cannot be made.

A related problem for any analysis is whether it can reject intermediate effects. The pulse is not a side effect, as are heart sounds, but an intermediate effect, linking the heart and the function, circulation. Since a heart cannot produce circulation without also producing a pulse, once again, necessary condition and requirement analyses are ruled out, and, because a pulse improves survival rate inasmuch as the organism would immediately die without it, a natural selection analysis fails. In contrast, intermediate effects in the causal chain leading to the functional effect pose no threat to an intentionalistic analysis. Just as the function of a fuel pump on an automobile is not to impart pulsations to the fuel simply because that is not an intended effect but merely a necessary effect on the way to the intended effect of supplying fuel to the engine, so also the pulse in humans is not a function of the heart because it is not an intended effect but merely a necessary effect on the way to the intended effect, circulation. The fact that the fuel pump analogy fits so naturally is, itself, suggestive.

The problem of how to accommodate malfunctions raised difficulties for the several analyses that required that functions always or usually succeed. On the intentionalistic analysis, because functions are the supposed or planned means in a means-end relation, an item may easily have as a function something it does not do. Thus, one might make a knife from soft iron, and its function may still be to cut wood even though it cannot do so. It is equally easy for a function to be something logically or physically impossible, as a computer program with the function of finding the terminating numeral in π or a rocket engine having the function of acceleration to superluminal velocity. The problem of contingent and necessary malfunctions is treated in the same way as the problem of missing and impossible goals, indicative of their close relation. Indeed, the concept of malfunction seems especially to support an intentionalistic analysis. One cannot explain a malfunction merely in terms of what an item fails to do, since everything fails to do an infinite number of things. In specifying which failure constitutes a malfunction, one inevitably talks about what something is *supposed* to do, and that is unequivocally intentional talk.

A related problem is that, although a function is normally beneficial to an organism, it sometimes becomes harmful, even deadly. In autoimmune diseases, such as multiple sclerosis or Guillain-Barré syndrome, the immune system, which normally attacks foreign cells, attacks host tissue instead. Due to cells malfunctioning, the host body is harmed. In some analyses, the problem of this kind of failure is addressed by means of a probability clause or by limiting functions to types. However, where the presence of the autoimmune disease becomes more widespread than its absence, probability and type strategies lead to error, requiring one to consider the destruction of host tissue as a function of the errant cells. The present analysis provides for any degree of harmful consequences because actual consequences need not match intended consequences.

Complexity apparently plays a role in our confidence in using teleological language. The complexity can be either in behavior, as in the behavior of spiders building webs or the division of labor of the social insects, or in structure, as in that of the ear or the eye or the intricate mounds of African termites that control temperature and humidity. There seems to be a little more confidence in saying that the function of the lens is to focus images on the retina than in saying, for example, that the function of the peppered moth's dark color, which resulted from industrial smoke in England, is protection from predators. Since complexity plays a role in distinguishing human artifacts from natural formations—it is more difficult, for example, to determine whether an item is or is not a knife than whether it is or is not a watch—this seems to be modest evidence that the use of teleological language is in some way linked to viewing things as intended.

At the other end of the scale, the fact that something can have a function without having any appreciable degree of complexity rules out the negative feedback analysis. Thus, as noted earlier, the function of fur on animals is to provide warmth, but the structure is too simple for negative feedback to be applicable.

Although it is common knowledge that teleological language has long been controversial when applied to lower forms of life and continues to be so, the significance of this fact goes quite unnoticed. If an analysis that successfully avoids mentalistic implications, such as one in terms of plasticity, necessary conditions, species survival, or negative feedback, were correct, it would be a mystery, since these analyses are not difficult to understand, why there is continued resistance and controversy regarding the use of teleological language in describing parts and processes of all organisms and the behavior of lower organisms and why such language is sometimes used within punctuation indicating nonstandard reading. An analysis could, of course, be correct but not recognized as such. Nevertheless, more is needed to explain why, after a debate as old as philosophy itself, the controversial status of teleological language continues, with each generation thinking the answers are at hand, only to pass the problems on to the next. This fact suggests that what is troublesome about teleological language is something inherent in it, something that cannot be removed without removing the very thing that makes it useful. An intentionalistic analysis does explain the enduring and profoundly controversial nature of teleological language and is the only analysis that does so. It accomplishes this precisely because of its most objectional feature, requiring a source for the external intention needed for the functions of parts and processes of organisms and for the goal-directed behavior of organisms judged incapable of being their own source of that intention. Nothing can be inferred about the nature of the external source required other than that it must be adequate for the effects and that seems to assume some ability.

The view that seems to meet the many conditions and restrictions of teleological language is the one that takes goals as fundamental to all teleological talk, links functions to goals by claiming functions to be a product of goal-directed behavior, and grounds both functions and goals on intentionality. This could be taken as reason enough to consign an intentionality analysis to oblivion. However, it should be remembered that one of the peculiar features of teleological language is that confidence in its use in describing goal-directed behavior varies roughly with the position on the phylogenetic scale of the organism, ranging from great confidence when used to describe human behavior to less or none when used to describe the behavior of lower animals and plants. An intentionality analysis is the only one that explains not only why teleological language remains controversial but also why there is a confidence scale governing its use and why that scale is the scale it is.

What to do about teleological language is a question quite different from what teleological language is and should be kept separate. Whether or not the evidence favors an intentionalistic analysis should be assessed on its own merits. If the intentionalistic analysis is correct, the options seem to be to admit an external agent into the worldview, thereby exceeding the limits of natural science, or to exclude the literal use of teleological language from the life sciences just as it has been excluded from the physical sciences. These choices are widely regarded as so extreme and untenable that any alternative is preferable. Where the stakes are this high, however, one must be alert to the danger that analyses that remove mind from teleology might be subjected to less than rigorous scrutiny, especially if there is the comfort of an alternative consensus. The problems posed by the use of teleological language in the life sciences have been around a long time and are genuine and deep.

Notes

1. Arthur Collins (1984, 357) says, "Actually, the words 'deliberate goals' are rather awkward. Why does not Nagel just say 'purposes and goals'? The answer is surely that Nagel's account of teleological explanation of nonhuman organic and machine functioning will speak of 'goals.'"

2. Since writing this, Woodfield has offered an analysis of intentionality in the context of philosophy of language (1990, 187-213).

3. However, Rosenberg (1986, 72-73) also seems to accept psychophysical identity. "This claim that human intentional states are not reducible to physical, neural states, nor specifiable by reference to a fixed type of behavior they result in, is not to deny that each and every particular cognitive state is identical to some particular neural state or other."

4. Other examples Boorse gives are the function of wood ash in the making of soap and the function of water in a Leyden jar in the electrical

experiments of Benjamin Franklin. The functions, also, he says, were unknown to the persons involved and so could not have been intended.

5. Thanks to Chris Wiggins, who influenced my thinking on this issue.

6. Clark Zumbach (1984, 8) takes a similar position: "Granted the microscopic mechanisms by which these changes take place were not and perhaps are still not known to many brewers. Nonetheless, the functional effects are known in a general way. It is accordingly unduly restrictive to claim that in these cases the brewer did not know the function of yeast in brewing. Thus, in spite of Boorse's objection, there is little reason to deny that intentions are necessarily a part of functional talk in the case of artifacts."

7. As noted earlier, Falk is also aware that teleological language is not extensional. He explains it without referring to minds by developing a theory of natural signs.

8. Philip Kitcher (1993, 379-397) takes the surprising view that a screw dropped accidentally into a machine, accidentally making a connection between two parts, thereby accidentally causing the machine to work, nevertheless has a function, even though the designer has no idea the connection is needed and did not intend that it be made. "Nevertheless, whatever satisfies that demand has the function of so doing" (381). He offers no argument for this unusual view, but uses it on which to build a theory of biological function in which there are natural selection pressures only at higher levels, such as the need to digest plants having cellulose. An item, say molars, may acquire a function merely by contributing to meeting such higher need, without being selected (382). "The functions of their constituents are understood in terms of the contributions made to the functioning of the whole" (382). This position, combining aspects of the theories of Wright and Cummins, is interesting, but, like Cummins', it accepts too much. It provides no way to block rain having the function of growing trees, since rain does contribute to trees surviving and flourishing, calcium having the function of growing bones, iron the function of producing hemoglobin, water the function of cooling the skin, and so on.

References

Achinstein, Peter. 1975[a]. Causation, Transparency, and Emphasis. *Canadian Journal of Philosophy* 5:1-23.

_____. 1975(b). Review of *The Philosophy of Biology* by Michael Ruse. *Canadian Journal of Philosophy* 4:745-54.

_____. 1977. Function Statements. *Philosophy of Science* 44:341-67.

_____. 1978. Teleology and Mentalism. *The Journal of Philosophy* 75:551-53.

_____. 1979. The Causal Relation. *Midwest Studies in Philosophy* 4: 369-86.

_____. 1983. Functional Explanation. Chap. 8 in *The Nature of Explanation*. Oxford: Oxford University Press.

Adams, Frederick R. 1979. A Goal-State Theory of Function Attributions. *Canadian Journal of Philosophy* 9:493-518.

Adams, Frederick R., and Berent Enç. 1988. Not Quite by Accident. *Dialogue* 27:287-97.

Allen, Colin, and Marc Bekoff. 1995. Biological Function, Adaptation, and Natural Design. *Philosophy of Science* 62: 609-22.

Amundson, Ron, and George V. Lauder. 1994. Function without Purpose: The Uses of Causal Role Function in Evolutionary Biology. *Biology and Philosophy* 9:443-69.

Anscombe, G. E. M. 1957. *Intention*. Oxford: Basil Blackwell.

Ayala, Francisco J. 1968. Biology as an Autonomous Science. *American Scientist* 56:207-21.

_____. 1970. Teleological Explanations in Evolutionary Biology. *Philosophy of Science* 37:1-15.

_____. 1977. Teleological Explanations. In *Evolution* by Theodosius Dobzhansky et al., 497-504. San Francisco: W. H. Freeman and Company. (Reprinted in *Philosophy of Biology*, ed. Michael Ruse, New York: Macmillan Publishing Company, 1989, 187-95.)

Baublys, Kenneth K. 1975. Comments on Some Recent Analyses of Functional Statements in Biology. *Philosophy of Science* 42:469-86.

Bechtel, William. 1983. Teleomechanism and the Strategy of Life. *Nature and System* 5:181-87.

_____. 1986. Teleological Functional Analyses and the Hierarchical Organization of Nature. In *Current Issues in Teleology*, ed. Nicholas Rescher, 26-48. *CPS Publications in Philosophy of Science*. Center for Philosophy of Science, University of Pittsburgh. Lanham, Md: University Press of America.

Beckner, Morton. 1959. *The Biological Way of Thought*. New York: Columbia University Press.

_____. 1969. Function and Teleology. *Journal of the History of Biology* 2:151-64. (Reprinted in *Topics in the Philosophy of Biology*, ed. Marjorie Grene and Everett Mendelsohn, vol. 27 of *Boston Studies in the Philosophy of Science*, Dordrecht: D. Reidel Publishing Company, 1976, 197-212.)

Bedau, Mark. 1990. Against Mentalism in Teleology. *American Philosophical Quarterly* 27:61-70.

_____. 1991. Can Biological Teleology Be Naturalized? *The Journal of Philosophy* 88:647-55.

_____. 1992[a]. Goal-Directed Systems and the Good. *The Monist* 75:34-51.

_____. 1992[b]. Where's the Good in Teleology? *Philosophy and Phenomenological Research* 52:781-806.

_____. 1993. Naturalism and Teleology. In *Naturalism: A Critical Appraisal*, ed. Steven J. Wagner and Richard Warner, 23-51. Notre Dame: University of Notre Dame Press.

Bennett, Jonathan. 1976. Teleology. Chap. 2 of *Linguistic Behaviour*, 36-81. Cambridge: Cambridge University Press.

_____. 1983. Teleology and Spinoza's Conatus. *Midwest Studies in Philosophy* 8:143-60.

Bernatowicz, A. J. 1958. Teleology in Science Teaching. *Science* 128:1402-405.

Bigelow, John, and Robert Pargetter. 1987. Functions. *The Journal of Philosophy* 84:181-96.

_____. 1990. Functional Explanation. Sec. 7.4 of chap. 7 in *Science and Necessity*, 323-41. Cambridge: Cambridge University Press.

Block, N. J. 1971. Are Mechanistic and Teleological Explanations of Behaviour Incompatible? *The Philosophical Quarterly* 21:109-17.

Boden, Margaret A. 1970. Intentionality and Physical Systems. *Philosophy of Science* 37:200-214.

_____. 1972. *Purposive Explanation in Psychology*. Cambridge: Harvard University Press.

Boorse, Christopher. 1976. Wright on Functions. *The Philosophical Review* 85:70-86.

Borger, Robert. 1970. Comment. In *Explanation in the Behavioural Sciences*, ed. Robert Borger and Frank Cioffi, 80-88. Cambridge: Cambridge University Press.

Boylan, Michael. 1981. Mechanism and Teleology in Aristotle's Biology. *Apeiron* 15:96-102.

_____. 1986. Monadic and Systemic Teleology. In *Current Issues in Teleology*, ed. Nicholas Rescher, 15-25. *CPS Publications in Philosophy of Science*. Center for Philosophy of Science, University of Pittsburgh. Lanham, Md.: University Press of America.

Bozonis, George A. 1973. Some Remarks on Mechanical Explanation in Biology. *Diotima* 1:61-80.

Braithwaite, Richard Bevan. 1946-47. Teleological Explanation. *Proceedings of the Aristotelian Society* 47:1-20.

_____. 1964. Causal and Teleological Explanation. Chap. 10 of *Scientific Explanation*. Cambridge: Cambridge University Press. (Reprinted in *Purpose in Nature*, ed. John V. Canfield, Englewood Cliffs, N. J.: Prentice-Hall, Inc., 1966, 27-47.)

Brand, Myles. 1986. Intentional Actions and Plans. *Midwest Studies in Philosophy* 10:213-30.

Brandon, Robert N. 1981. Biological Teleology: Questions and Explanations. *Studies in History and Philosophy of Science* 12:91-105.

Bratman, Michael E. 1987. *Intentions, Plans, and Practical Reason*. Cambridge: Harvard University Press.

Broad, Charles Dunbar. 1918-19. Mechanical Explanation and its Alternatives. *Proceedings of the Aristotelian Society* 19:86-124.

Brodbeck, May. 1963. Meaning and Action. *Philosophy of Science* 30:309-24.

Brody, Baruch. 1975. The Reduction of Teleological Sciences. *American Philosophical Quarterly* 12:69-76.

Büchel, Wolfgang. 1982. Teleologie und Negentropie. *Zeitschrift für Allgemeine Wissenschaftstheorie* 13:40-47.

Burch, Robert W. 1978. Functional Explanation and Normalcy. *Southwestern Journal of Philosophy* 9:45-53.

Burks, Arthur W. 1988. Teleology and Logical Mechanism. *Synthese* 76:333-70.

Butts, Robert E. 1984. *Kant and the Double Government Methodology.* Dordrecht: D. Reidel Publishing Company.

Byerly, Henry. 1979. Teleology and Evolutionary Theory: Mechanisms and Meanings. *Nature and System* 1:157-76.

Canfield, John V. 1964. Teleological Explanation in Biology. *The British Journal for the Philosophy of Science* 14:285-95.

_____. 1965. Teleological Explanation in Biology: A Reply. *The British Journal for the Philosophy of Science* 15:327-31.

_____, ed. 1966. *Purpose in Nature.* Englewood Cliffs, N. J.: Prentice-Hall, Inc.

_____. 1978. Review of *Teleological Explanations* by Larry Wright and *Teleology* by Andrew Woodfield. *The Philosophical Review* 87:284-88.

_____. 1990. The Concept of Function in Biology. *Philosophical Topics* 18:29-53.

Cartwright, Nancy. 1986. Two Kinds of Teleological Explanation. In *Human Nature and Natural Knowledge*, ed. Alan Donagan, Anthony N. Perovich, Jr., and Michael I. Wedin, 201-10. Vol. 89 of *Boston Studies in the Philosophy of Science.* Dordrecht: D. Reidel Publishing Company.

Cellerier, Guy. 1983. The Historical Genesis of Cybernetics: Is Teleonomy a Category of Understanding? *Nature and System* 5:211-25.

Centore, F. F. 1972. Mechanism, Teleology, and 17th Century English Science. *International Philosophical Quarterly* 12:553-71.

Cohen, G. A. 1982. Functional Explanation, Consequence Explanation, and Marxism. *Inquiry* 25:27-56.

Cohen, Jonathan. 1951. Teleological Explanation. *Proceedings of the Aristotelian Society* 51:255-92. (Reprinted in *Purposive Behaviour and Teleological Explanations*, ed. Frank George and Les Johnson, New York: Gordon and Breach Science Publishers, 1985, 40-68.)

Collin, Finn. 1987. Natural Kind Terms and Explanation of Human Action. Vol. 5, part 2, sec. 11 of *Abstracts, 8th International Congress of Logic, Methodology and Philosophy of Science*, comp. Dr. V. L. Rabinovich, 349-51. Institute of Philosophy of the Academy of Sciences of the USSR. Moscow, USSR.

Collins, Arthur W. 1978. Teleological Reasoning. *The Journal of Philosophy* 75:540-50.

_____. 1984. Action, Causality, and Teleological Explanation. *Midwest Studies in Philosophy* 9:345-69.

Cowan, J. L. 1968. Purpose and Teleology. *The Monist* 52:317-28.

Cummins, Robert. 1975. Functional Analysis. *The Journal of Philosophy* 72:741-65. (Reprinted in *Conceptual Issues in Evolutionary Biology*, ed. Elliot Sober. Cambridge: The MIT Press, 1994, 49-69.)

Darwin, Charles. 1878. *The Origin of Species by Means of Natural Selection*. New York: D. Appleton and Company.

Davies, Paul Sheldon. 1994. Troubles for Direct Proper Functions. *Nous* 28:363-81.

_____. 1995. "Defending" Direct Proper Functions. *Analysis* 55:299-306.

Downes, Chauncey. 1976. Functional Explanations and Intentions. *Philosophy of the Social Sciences* 6:215-25.

Dretske, Fred. 1986[a]. Aspects of Cognitive Representation. In *The Representation of Knowledge and Belief*, ed. Myles Brand and Robert M. Harnish, 101-15. Tucson: The University of Arizona Press.

_____. 1986[b]. Misrepresentation. In *Belief: Form, Content, and Function*, ed. Radu J. Bogdan, 17-36. Oxford: Clarendon Press.

_____. 1990. Reply to Reviewers. *Philosophy and Phenomenological Research* 50:819-39.

_____. 1992. *Explaining Behavior: Reasons in a World of Causes*. Cambridge: The MIT Press.

_____. 1993. The Nature of Thought. *Philosophical Studies* 70:185-99.

Ducasse, C. J. 1925. Explanation, Mechanism and Teleology. *The Journal of Philosophy* 22:150-55. (Reprinted in *Purposive Behaviour and Teleological Explanations*, ed. Frank George and Les Johnson, New York: Gordon and Breach Science Publishers, 1985, 35-39; and in *Readings in Philosophical Analysis*, ed. Herbert Feigl and Wilfrid Sellars, New York: Appleton-Century-Crofts, 1949, 540-44.)

Duchesneau, François. 1977. Analyse Fonctionelle et Principe des Conditions d'Existence Biologique. *Revue Internationale de Philosophie* 31:285-312.

_____. 1978. Téléologie et Détermination Positive de l'Ordre Biologique. *Dialectica* 32:135-53.

Ehring, Douglas. 1983. Goal-Directed Processes. *Southwest Philosophical Studies* 9:39-47.

_____. 1984[a]. Negative Feedback and Goals. *Nature and System* 6:217-20.

_____. 1984[b]. The System-Property Theory of Goal-Directed Processes. *Philosophy of the Social Sciences* 14:497-504.

_____. 1985. Enç on Functions. *Philosophical Inquiry* 7:74-81.

_____. 1986[a]. Accidental Functions. *Dialogue* 25:291-302.

_____. 1986[b]. Teleology and Impossible Goals. *Philosophy and Phenomenological Research* 47:127-31.

Elder, Crawford L. 1994. Proper Functions Defended. *Analysis* 54:167-71.

Enç, Berent. 1979. Function Attributions and Functional Explanations. *Philosophy of Science* 46:343-65.

Enç, Berent, and Fred Adams. 1992. Functions and Goal Directedness. *Philosophy of Science* 59:635-54.

Engels, Eve-Marie. 1978. Teleologie—Eine "Sache der Formulierung" oder eine "Formulierung der Sache"? Überlegungen zu Ernest Nagels Reduktionistischer Strategie und Versuch ihrer Widerlegung. *Zeitschrift für Allgemeine Wissenschaftheorie* 9:225-35.

_____, 1982. *Die Teleologie des Lebendigen: Eine Historisch-Systematische Untersuchung.* Vol. 63 of *Erfahrung und Denken: Schriften zür Forderung der Beziehungen zwischen Philosophie und Einselwissenschaften.* Berlin: Duncker and Humblot.

Esposito, Joseph L. 1980-81. Teleological Causation. *The Philosophical Forum* 12:116-27.

Faber, Roger J. 1984. Feedback, Selection, and Function: A Reductionistic Account of Goal-Orientation. In *Methodology, Metaphysics and the History of Science*, ed. Robert S. Cohen and Marx W. Wartofsky, 43-135. Vol. 84 of *Boston Studies in the Philosophy of Science*. Dordrecht: D. Reidel Publishing Company.

_____. 1986. *Clockwork Garden: On the Mechanistic Reduction of Living Things.* Amherst: University of Massachusetts Press.

Falk, Arthur E. 1981. Purpose, Feedback, and Evolution. *Philosophy of Science* 48:198-217. (Reprinted in *Purposive Behaviour and Teleological Explanations*, ed. Frank George and Les Johnson, New York: Gordon and Breach Science Publishers, 1985, 291-309.)

_____. 1990. Teleology and Nature's Semeiotics: From Aristotle to Feedback, by Way of Kant. (Unpublished paper.)

_____. 1995. Essay on Nature's Semeiosis. *Journal of Philosophical Research* 20:298-348.

Frankfurt, Harry G., and Brian Poole. 1966. Functional Analyses in Biology. *The British Journal for the Philosophy of Science* 17:69-72.

Frolov, I. T. 1977. Organic Determinism and Teleology in Biological Research. In *Foundational Problems in the Special Sciences*, ed. Robert E. Butts and Jaakko Hintikka, 119-29. Part 2 of *Proceedings of the 5th International Congress of Logic, Methodology and Philosophy of Science, London, Ontario, Canada, 1975.* Vol. 10 of *University of*

Western Ontario Series in Philosophy of Science. Dordrecht: D. Reidel Publishing Company.

Geach, Peter. 1975[a]. Teleological Explanation. In *Explanation*, ed. Stephen Körner, 76-95. New Haven: Yale University Press.

_____. 1975[b]. Reply to Comments. In *Explanation*, ed. Stephen Körner, 112-17. New Haven: Yale University Press.

George, Frank, and Les Johnson, eds. 1985. *Purposive Behaviour and Teleological Explanations*. Vol. 8 of *Studies in Cybernetics*. New York: Gordon and Breach Science Publishers.

Godfrey-Smith, Peter. 1993. Functions: Consensus without Unity. *Pacific Philosophical Quarterly* 74:196-208.

_____. 1994. A Modern History Theory of Function. *Nous* 28:344-62.

_____. 1996. *Complexity and the Function of Mind in Nature*. Cambridge: Cambridge University Press.

Goldman, Alvin I. 1970. *A Theory of Action*. Princeton, N.J.: Princeton University Press.

Goldstein, Leon J. 1976. Review of *Teleological Explanation* by Larry Wright. *International Studies in Philosophy* 8:191-92.

Greenstein, Harold. 1973. The Logic of Functional Explanations. *Philosophia* 3:247-64.

Griffiths, Paul E. 1993. Functional Analysis and Proper Functions. *The British Journal for the Philosophy of Science* 44:409-22.

Grim, Patrick. 1974. Wright on Functions. *Analysis* 35:62-64.

_____. 1977. Further Notes on Functions. *Analysis* 37:169-76.

Gruner, Rolf. 1966. Teleological and Functional Explanations. *Mind* 75:516-26.

Hall, Richard J. 1990. Does Representational Content Arise from Biological Function? Vol. 1 of *PSA 1990: Proceedings of the 1990 Biennial Meeting of the Philosophy of Science Association*, ed. Arthur Fine, Micky Forbes, and Linda Wessels, 193-99. East Lansing, Mich.: Philosophy of Science Association.

Harris, Errol E. 1973. Mechanism and Teleology in Contemporary Thought. *Philosophy in Context* 2:49-55.

Hausman, Daniel M. 1993. Linking Causal and Explanatory Asymmetry. *Philosophy of Science* 60:435-51.

Hausman, David B. 1978. The Paradox of Teleological Ascription. *The Journal of Medicine and Philosophy* 3:144-57.

_____. 1985. The Explanation of Goal-Directed Behavior. *Synthese* 65:327-46.

Hempel, Carl, G. 1948. Studies in the Logic of Explanation. *Philosophy of Science* 15:135-75. (Reprinted in *Aspects of Scientific Explanation and Other Essays in the Philosophy of Science* by Carl G. Hempel, New York: The Free Press, 1965, 245-90.)

_____. 1959. The Logic of Functional Analysis. In *Symposium on Sociological Theory*, ed. Llewellyn Gross, 271-307. New York: Harper and Row. (Reprinted in *Aspects of Scientific Explanation and Other Essays in the Philosophy of Science* by Carl G. Hempel, New York: The Free Press, 1965, 297-330; and in *Purpose in Nature*, ed. John V. Canfield, Englewood Cliffs, N. J.: Prentice-Hall, Inc., 1966. 89-108.)

_____. 1962. Explanation in Science and in History. In *Frontiers of Science and Philosophy*, ed. Robert G. Colodny, 7-33. Pittsburgh: University of Pittsburgh Press.

_____. 1965. *Aspects of Scientific Explanation and Other Essays in the Philosophy of Science*. New York: The Free Press.

Henry, Grete. 1975. Comment. In *Explanations*, ed. Stephen Körner, 105-12. New Haven: Yale University Press.

Hirschman, David. 1973. Function and Explanation, Part I. *Proceedings of the Aristotelian Society*. Supplementary vol. 47:19-38.

Horan, Barbara L. 1989[a]. Functional Explanations in Sociobiology. *Biology and Philosophy* 4:131-58.

_____. 1989[b]. Functional Explanations in Sociobiology: A Reply to Critics. *Biology and Philosophy* 4:205-28.

Hull, David L. 1973. A Belated Reply to Gruner. *Mind* 82:437-38.

_____. 1974. *Philosophy of Biological Science*. Englewood Cliffs, N. J.: Prentice-Hall, Inc.

_____. 1979. Philosophy of Biology. In *Current Research in Philosophy of Science: Proceedings of the P.S.A. Critical Research Problems Conference*, ed. Peter D. Asquith and Henry E. Kyburg, Jr., 421-35. East Lansing, Mich.: Philosophy of Science Association.

_____. 1982. Philosophy and Biology. In *Philosophy of Science*, ed. Guttorm Fløistad, 281-316. Vol. 2 of *Contemporary Philosophy: A New Survey*. The Hague: Martinus Nijhoff.

Jacobs, Jonathan. 1986[a]. Teleology and Reduction in Biology. *Biology and Philosophy* 1:389-99.

_____. 1986[b]. Teleological Form and Explanation. In *Current Issues in Teleology*, ed. Nicholas Rescher, 49-55. *CPS Publications in Philosophy of Science*. Center for Philosophy of Science, University of Pittsburgh. Lanham, Md.: University Press of America.

Kernohan, Andrew. 1987. Teleology and Logical Form. *The British Journal for the Philosophy of Science* 38:27-34.

Kitchener, Richard F. 1976. On Translating Teleological Explanations. *International Logic Review* 7:50-55.

Kitcher, Philip. 1993. Function and Design. *Midwest Studies in Philosophy* 18:379-97.

Kleiner, Scott A. 1975. Essay Review: The Philosophy of Biology. *Southern Journal of Philosophy* 13:523-42.

Kraemer, Eric Russert. 1979. On the Causal Irreducibility of Natural Function Statements. In *Transactions of the Nebraska Academy of Sciences* 7:149-52.

_____. 1984[a]. Consciousness and the Exclusivity of Function. *Mind* 93:271-75.

_____. 1984.[b] Teleology and Organism-Body Problem. *Metaphilosophy* 15:45-54.

_____. 1985. Function, Law and the Explanation of Intentionality. In *Philosophy of Mind, Philosophy of Psychology: Proceedings of the 9th International Wittgenstein Symposium, 19th to 26th August 1984, Kirchberg/Wechsel, Austria*, 90-93. Vienna: Hölder-Pichler-Tempsky.

Krikorian, Y. H. 1948. Teleology and Causality. *The Review of Metaphysics* 2:35-46.

Langford, Glenn. 1981. The Nature of Purpose. *Mind* 90:1-19.

Latta, R. 1908. Purpose. *Proceedings of the Aristotelian Society* 8:17-32.

Lehman, Hugh. 1965[a]. Functional Explanation in Biology. *Philosophy of Science* 32:1-20.

_____. 1965[b]. Teleological Explanation in Biology. *The British Journal for the Philosophy of Science* 15:327.

Lenoir, Timothy. 1981. Teleology without Regrets: The Transformation of Physiology in Germany: 1790-1847. *Studies in History and Philosophy of Science* 12:293-54.

Levin, Michael. 1976. On the Ascription of Functions to Objects, with Special Reference to Inference in Archaeology. *Philosophy of the Social Sciences* 6:227-34.

Machamer, Peter. 1977. Teleology and Selective Processes. In *Logic, Laws and Life: Some Philosophical Complications*, ed. Robert G. Colodny, 129-42. Vol. 6 of *University of Pittsburgh Series in the Philosophy of Science*. Pittsburgh: University of Pittsburgh Press.

MacIntyre, Alasdair. 1960. Purpose and Intelligent Action: I. *Proceedings of the Aristotelian Society* 34:79-96.

Mandelbaum, Maurice. 1982. G. A. Cohen's Defense of Functional Explanation. *Philosophy of the Social Sciences* 12:285-87.

Manier, Edward. 1969. "Fitness" and Some Explanatory Patterns in Biology. *Synthese* 20:206-18.

_____. 1971. Functionalism and the Negative Feedback Model in Biology. *PSA 1970: Proceedings of the 1970 Biennial Meeting, Philosophy of Science Association*, ed. Roger C. Buck and Robert S. Cohen, 225-40. Vol. 8 of *Boston Studies in the Philosophy of Science*. Dordrecht: D. Reidel Publishing Company.

Manning, Richard N. 1997. Biological Function, Selection, and Reduction. *The British Journal for the Philosophy of Science* 48: 69-82.

Manser, A. R. 1973. Function and Explanation: Part II. *Proceedings of the Aristotelian Society* 47:39-52.

Margolis, Joseph. 1968. Taylor on the Reduction of Teleological Laws. *Inquiry* 11:118-24.

Matthen, Mohan. 1988. Biological Functions and Perceptual Content. *The Journal of Philosophy* 85:5-27.

_____. 1991. Naturalism and Teleology. *The Journal of Philosophy* 88:656-57.

Matthen, Mohan, and Edwin Levy. 1984. Teleology, Error, and the Human Immune System. *The Journal of Philosophy* 81:351-72.

_____. 1986. Organic Teleology. In *Current Issues in Teleology*, ed. Nicholas Rescher, 93-101. *CPS Publications in Philosophy of Science*. Center for Philosophy of Science, University of Pittsburgh. Lanham, Md.: University Press of America.

Mayr, Ernst. 1974. Teleological and Teleonomic, a New Analysis. In *Methodological and Historical Essays in the Natural and Social Sciences*, ed. R. S. Cohen and M. W. Wartofsky, 91-117. Vol. 14 of *Boston Studies in the Philosophy of Science*. Dordrecht: D. Reidel Publishing Company. (Reprinted in *Toward a New Philosophy of Biology*, ed. Ernst Mayr, Cambridge: Harvard University Press, 1988, 38-66.)

McFarland, John D. 1970. *Kant's Concept of Teleology*. Edinburgh: University of Edinburgh Press.

McLachlan, Hugh V. 1976. Functionalism, Causation and Explanation. *Philosophy of the Social Sciences* 6:235-40.

Melander, Peter. 1993. How Not to Explain the Errors of the Immune System. *Philosophy of Science* 60:223-41.

Millikan, Ruth Garrett. 1984. *Language, Thought, and Other Biological Categories: New Foundations for Realism*. Cambridge: The MIT Press.

_____. 1989[a]. An Ambiguity in the Notion "Function." *Biology and Philosophy* 4:172-76.

_____. 1989[b]. Biosemantics. *The Journal of Philosophy* 86:281-97. (Reprinted in *White Queen Psychology and Other Essays for Alice* by Ruth Garrett Millikan, Cambridge: The MIT Press, 1993, 83-101.)

_____. 1989[c]. In Defense of Proper Functions. *Philosophy of Science* 56:288-302. (Reprinted in *White Queen Psychology and Other Essays for Alice* by Ruth Garrett Millikan, Cambridge: The MIT Press, 1993, 13-29.)

_____. 1990[a]. Compare and Contrast Dretske, Fodor, and Millikan on Teleosemantics. *Philosophical Topics* 18:151-61. (Reprinted in *White Queen Psychology and Other Essays for Alice* by Ruth Garrett Millikan, Cambridge: The MIT Press, 1993, 123-33.)

_____. 1990[b]. Seismograph Readings for "Explaining Behavior." *Philosophy and Phenomenological Research* 50:807-12.

_____. 1993[a]. Explanation in Biopsychology. In *Mental Causation*, ed. John Heil and Alfred Mele, 211-32. Oxford: Clarendon Press. (Reprinted in *White Queen Psychology and Other Essays for Alice* by Ruth Garrett Millikan, Cambridge: The MIT Press, 1993, 171-92.)

_____. 1993[b]. *White Queen Psychology and Other Essays for Alice.* Cambridge: The MIT Press.

Minton, Arthur J. 1975. Wright and Taylor: Empiricist Teleology. *Philosophy of Science* 42:299-306.

Mitchell, Sandra D. 1989. The Causal Background of Functional Explanation. *International Studies in the Philosophy of Science: The Dubrovnik Papers* 3:213-29.

_____. 1995. Function, Fitness and Disposition. *Biology and Philosophy* 10:39-54.

Montefiore, Alan. 1971. Final Causes: II. *Proceedings of the Aristotelian Society*. Supplementary vol. 45:171-92.

Munson, Ronald. 1971. Biological Adaptation. *Philosophy of Science* 38:200-15.

_____. 1972. Biological Adaptation: A Reply. *Philosophy of Science* 39:529-32.

Nagel, Ernest. 1961. The Structure of Teleological Explanations. Chap. 12, part 1 of *The Structure of Science: Problems in the Logic of Scientific Explanation*. New York and Burlingame: Harcourt, Brace & World, 401-28. (Reprinted in *Purpose in Nature*, ed. John V. Canfield, Englewood Cliffs, N.J.: Prentice-Hall, Inc., 67-88.)

_____. 1977. Teleology Revisited. *The Journal of Philosophy* 74:261-301. (Reprinted in *Teleology Revisited and Other Essays in the Philosophy and History of Science* by Ernest Nagel, New York: Columbia University Press, 1979, 275-316.)

Neander, Karen. 1988. What Does Natural Selection Explain? Correction to Sober. *Philosophy of Science* 55:422-26.

_____. 1991[a]. Functions as Selected Effects: The Conceptual Analyst's Defense. *Philosophy of Science* 58:168-84.

_____. 1991[b]. The Teleological Notion of "Function." *Australasian Journal of Philosophy* 69:454-68.

Nissen, Lowell. 1970. Canfield's Functional Translation Schema. *The British Journal for the Philosophy of Science* 21:193-95.

_____. 1971. Neutral Functional Statement Schemata *Philosophy of Science* 38:251-57. (Reprinted in *Purposive Behaviour and Teleological Explanations*, ed. Frank George and Les Johnson, New York: Gordon and Breach Science Publishers, 1985, 260-67.)

_____. 1977. Wimsatt on Function Statements. *Studies in History and Philosophy of Science* 8:341-47.

_____. 1979[a]. Review of *Teleological Explanations: An Etiological Analysis of Goals and Functions* by Larry Wright. *The Thomist* 43:337-41.

_____. 1979[b]. Review of *Teleology* by Andrew Woodfield. *The Thomist* 43:341-47.

_____. 1980-81. Nagel's Self-Regulation Analysis of Teleology. *The Philosophical Forum* 12:128-38.

_____. 1981. Wright's Teleological Analysis versus Impossible Goals. *Proceedings of the Southwestern Philosophical Society. Philosophical Topics*. Supplementary vol. 12:125-31.

_____. 1983[a]. Wright and Woodfield on Natural Functions and Reverse Causation. Vol. 4, sec. 9 of *Abstracts of the 7th International Congress of Logic, Methodology and Philosophy of Science*, comp. Paul Weingartner, 321-24. Salzburg, Austria.

_____. 1983[b]. Wright on Teleological Descriptions of Goal-Directed Behavior. *Philosophy of Science* 50:151-58.

_____. 1984. Woodfield's Analysis of Teleology. *Philosophy of Science* 51:488-94.

_____. 1986. Natural Functions and Reverse Causation. *Current Issues in Teleology*, ed. Nicholas Rescher, 129-35. *CPS Publications in Philosophy of Science*. Center for Philosophy of Science, University of Pittsburgh. Lanham, Md.: University Press of America.

_____. 1987. Three Ways of Eliminating Mind from Teleology. Vol. 2, sec. 9 of *Abstracts of 8th International Congress of Logic, Methodology and Philosophy of Science*, comp. Dr. V. L. Rabinovich, 282-85. Institute of Philosophy of the Academy of Sciences of the USSR. Moscow, USSR.

_____. 1991. Teleology and Natural Selection. Vol. 3, sec. 12 of *Abstracts of the 9th International Congress of Logic, Methodology and Philosophy of Science*, 65. Uppsala: Sweden, Uppsala University.

_____. 1993. Four Ways of Eliminating Mind from Teleology. *Studies in History and Philosophy of Science* 24:27-48.

Noble, Denis. 1966-67. Charles Taylor on Teleological Explanation. *Analysis* 27:96-103.

_____. 1967-68. The Conceptual View of Teleology. *Analysis* 28:62-63.

Nowell-Smith, P. H. 1960. Purpose and Intelligent Actions: II. *Proceedings of the Aristotelian Society*. Supplementary vol. 34:97-112.

Papineau, David. 1984. Representation and Explanation. *Philosophy of Science* 51:550-72.

_____. 1987. *Reality and Representation*. Oxford: Basil Blackwell.

_____. 1990. Truth and Teleology. In *Explanation and its Limits*, ed. Dudley Knowles, 21-43. Cambridge: Cambridge University Press.

_____. 1991. Teleology and Mental States: II. *Proceedings of the Aristotelian Society*. Supplementary vol. 65:33-54.

Pfeiffer, Alfred. 1974. Causalité et Finalité *Deutsche Zeitschrift für Philosophie* 22:83-85.

Plantinga, Alvin. 1988. Positive Epistemic Status and Proper Function. *Philosophical Perspectives* 2:1-50.

_____. 1993. *Warrant and Proper Function*. Oxford: Oxford University Press.

Porpora, Douglas V. 1980. Operant Conditioning and Teleology. *Philosophy of Science* 47:568-82.

Pratt, Vernon. 1978. Review of Woodfield's *Teleology*. *Mind* 87:312-14.

Prior, Elizabeth W. 1985. What is Wrong with Etiological Accounts of Biological Function? *Pacific Philosophical Quarterly* 66:310-28.

Purton, A. C. 1979. Biological Function. *The Philosophical Quarterly* 29:10-24.

Resnik, David B. 1995. Functional Language and Biological Discovery. *Journal for General Philosophy of Science* 26:119-34.

Richmond, Samuel A. 1973. Comments on "Mechanism and Teleology in Contemporary Thought" II. *Philosophy in Context* 2:59-61.

Rignano, Eugenio. 1931. The Concept of Purpose in Biology. *Mind* 40:335-40.

Ringen, Jon D. 1976. Explanation, Teleology, and Operant Behaviorism: A Study of the Experimental Analysis of Purposive Behavior. *Philosophy of Science* 43:223-53.

_____. 1985. Operant Conditioning and a Paradox of Teleology. *Philosophy of Science* 52:565-77.

Roll-Hanson, Nils. 1976. Critical Teleology: Immanual Kant and Claude Bernard on the Limitations of Experimental Biology. *Journal of the History of Biology* 9:59-91.

Root, Michael. 1989. Covering Laws and Functions. *Biology and Philosophy* 4:185-90.

Roque, Alicia Juarrero. 1981. Dispositions, Teleology and Reductionism. *Philosophical Topics* 12:153-65.

Ros, Arno. 1982. Kausale, Teleologische und Teleonomische Erklärungen. *Zeitschrift für Allgemeine Wissenschaftstheorie* 13:320-35.

Rosenberg, Alexander. 1982. Causation and Teleology in Contemporary Philosophy of Science. In *Philosophy of Science*, ed. Guttorm Fløistad, 51-86. Vol. 2 of *Contemporary Philosophy: A New Survey*. The Hague: Martinus Nijhoff.

_____. 1985. Teleology and the Roots of Autonomy. Chap. 3 of *The Structure of Biological Science*, 37-68. Cambridge: Cambridge University Press.

_____. 1986. Intention and Action among the Macromolecules. *Current Issues in Teleology*, ed. Nicholas Rescher, 65-76. *CPS Publications in Philosophy of Science*. Center for Philosophy of Science, University of Pittsburgh. Lanham, Md.: University Press of America.

_____. 1989. Perceptual Presentations and Biological Function: A Comment on Matthen. *The Journal of Philosophy* 86:38-44.

Rosenblueth, Arturo, and Norbert Wiener. 1950. Purposeful and Non-Purposeful Behavior. *Philosophy of Science* 17:318-26. (Reprinted in *Purposive Behaviour and Teleological Explanations*, ed. Frank George and Les Johnson, New York: Gordon and Breach Science Publishers, 1985, 18-27.)

Rosenblueth, Arturo, Norbert Wiener, and Julian Bigelow. 1943. Behavior, Purpose and Teleology. *Philosophy of Science* 10:18-24. (Reprinted in *Purpose in Nature*, ed. John V. Canfield, Englewood Cliffs, N.J.: Prentice-Hall, Inc., 1966, 9-16; and in *Purposive Behaviour and Teleological Explanations*, ed. Frank George and Les Johnson, New York: Gordon and Breach Science Publishers, 1985, 1-8.)

Rudner, Richard S. 1966. Functionalism and Other Problems of Teleological Inquiry. Chap. 5 of *Philosophy of the Social Sciences*. Englewood Cliffs, N.J.: Prentice-Hall, Inc., 84-111.

Ruse, Michael. 1971. Functional Statements in Biology. *Philosophy of Science* 38:87-95.

_____. 1972. Biological Adaptation. *Philosophy of Science* 39:525-28.

_____. 1973[a]. *The Philosophy of Biology*. London: Hutchinson & Co. Ltd.

_____. 1973[b]. A Reply to Wright's Analysis of Functional Statements. *Philosophy of Science* 40:277-80.

_____. 1973[c]. Teleological Explanation and the Animal World. *Mind* 82:433-36.

_____. 1977. Is Biology Different from Physics? In *Logic, Laws, and Life: Some Philosophical Complications*, ed. Robert G. Colodny, 89-127. Vol. 6 of *University of Pittsburgh Series in the Philosophy of Science*. Pittsburgh: University of Pittsburgh Press.

_____. 1978. Review of *Teleology* by Andrew Woodfield and *Teleological Explanations* by Larry Wright. *Canadian Journal of Philosophy* 8:191-203.

_____. 1979. Philosophy of Biology Today: No Grounds for Complacency. *Philosophia* 8:785-96.

_____. 1982. Teleology Redux. In *Scientific Philosophy Today: Essays in Honor of Mario Bunge*, ed. Joseph Agassi and Robert S. Cohen, 299-309. Vol. 67 of *Boston Studies in the Philosophy of Science*. Dordrecht: D. Reidel Publishing Company.

_____. 1986. Teleology and the Biological Sciences. *Current Issues in Teleology*, ed. Nicholas Rescher, 56-64. *CPS Publications in Philosophy of Science*. Center for Philosophy of Science, University of Pittsburgh. Lanham: Md.: University Press of America.

_____. ed. 1989. *Philosophy of Biology*. New York: Macmillan Publishing Company.

Russell, E. S. 1922-23. Psychobiology. *Proceedings of the Aristotelian Society* 23:141-56.

_____. 1932-33. The Limitations of Analysis in Biology. *Proceedings of the Aristotelian Society* 33:147-58.

_____. 1945. *The Directiveness of Organic Activities*, Cambridge: Cambridge University Press.

Salmon, Merrilee H. 1981. Ascribing Functions to Archaeological Objects. *Philosophy of the Social Sciences* 11:19-25.

Scheffler, Israel. 1959. Thoughts on Teleology. *The British Journal for the Philosophy of Science* 9:265-84. (Reprinted in *Purpose in Nature*, ed. John V. Canfield, Englewood Cliffs, N. J.: Prentice-Hall, Inc., 1966, 48-66; and in *Purposive Behaviour and Teleological Explanations*, ed. Frank George and Les Johnson, New York: Gordon and Breach Science Publishers, 1985, 69-87.)

_____. 1963. *The Anatomy of Inquiry*. New York. Alfred A. Knopf.

Shelanski, Vivian B. 1973. Nagel's Translation of Teleological Statements: A Critique. *The British Journal for the Philosophy of Science* 24:397-401.

Sher, George. 1975. Charles Taylor on Purpose and Causation. *Theory and Decision* 6:27-38.

Short, T. L. 1981. Peirce's Concept of Final Causation. *Transactions of the Charles S. Peirce Society* 17:369-82.

_____. 1983. Teleology in Nature. *American Philosophical Quarterly* 20:311-20.

Shrader-Frechette, Kristin. 1986. Organismic Biology and Ecosystems Ecology: Description or Explanation? *Current Issues in Teleology*, ed. Nicholas Rescher, 77-92. *CPS Publications in Philosophy of Science*. Center for Philosophy of Science, University of Pittsburgh. Lanham: Md.: University Press of America.

Simon, Michael A. 1971. *The Matter of Life: Philosophical Problems of Biology*. New Haven: Yale University Press.

Simon, Thomas W. 1976. A Cybernetic Analysis of Goal-Directedness. Vol. 1 of *PSA 1976: Proceedings of the 1976 Biennial Meeting of the Philosophy of Science Association*, ed. Frederick Suppe and Peter D. Asquith, 56-67. East Lansing, Mich.: Philosophy of Science Association.

Smart, J. J. C. 1968. Explanation in Biological Sciences. Chap. 4 of *Between Science and Philosophy: An Introduction to the Philosophy of Science*, 91-120. New York: Random House.

Sober, Elliot, ed. 1994. *Conceptual Issues in Evolutionary Biology*. Cambridge: The MIT Press.

Sommerhoff, Gerd. 1950. *Analytical Biology*. Oxford: Oxford University Press.

Sorabji, Richard. 1964. Function. *The Philosophical Quarterly* 14:289-302.

Spaemann, Robert. 1978. Naturteologie und Handlung. *Zeitschrift für Philosophische Forschung* 32:481-93.

Sprigge, Timothy L. S. 1971. Final Causes, Part I. *Proceedings of the Aristotelian Society.* Supplementary vol. 45:149-70.

Stegmüller, Wolfgang. 1969. Teleologie, Funktionalanalyse und Selbstregulation. Chap. 8 of *Wissenschaftliche Erklärung und Begründung*, 518-623. Vol. 1 of *Probleme und Resultate der Wissenschaftstheorie und Analytischen Philosophie*. Berlin: Springer-Verlag.

Sutherland, N. S. 1970. Is the Brain a Physical System? In *Explanation in the Behavioural Sciences*, ed. Robert Borger and Frank Cioffi, 97-122. Cambridge: Cambridge University Press.

Taylor, Charles. 1964. *The Explanation of Behaviour*. New York: Humanities Press. (Chap. 1, Purpose and Teleology, is reprinted in *Purposive Behaviour and Teleological Explanations*, ed. Frank George and Les Johnson, New York: Gordon and Breach Science Publishers, 1985, 88-106.)

_____. 1966-67. Teleological Explanation—A Reply to Denis Noble. *Analysis* 27:141-43. (Reprinted in *Purposive Behaviour and Teleological Explanations*, ed. Frank George and Les Johnson, New York: Gordon and Breach Science Publishers, 1985, 107-109.)

_____. 1968. Reply to Margolis. *Inquiry* 11:124-28.

_____. 1970[a]. The Explanation of Purposive Behaviour. In *Explanation in the Behavioural Sciences*, ed. Robert Borger and Frank Cioffi. Cambridge: Cambridge University Press, 49-79.

_____. 1970[b]. Reply. In *Explanation in the Behavioural Sciences*, ed. Robert Borger and Frank Cioffi. Cambridge: Cambridge University Press, 89-95.

Taylor, Richard. 1950[a]. Comments on a Mechanistic Conception of Purposefulness. *Philosophy of Science* 17:310-17. (Reprinted in *Purpose in Nature*, ed. John V. Canfield, Englewood Cliffs, N. J.: Prentice-Hall, Inc., 1966, 17-26, and in *Purposive Behaviour and Teleological Explanations*, ed. Frank George and Les Johnson, New York: Gordon and Breach Science Publishers, 1985, 9-17.)

_____. 1950[b]. Purposeful and Non-Purposeful Behavior: a Rejoinder. *Philosophy of Science* 17:327-32. (Reprinted in *Purposive Behaviour and Teleological Explanations*, ed. Frank George and Les Johnson, New York: Gordon and Breach Science Publishers, 1985, 28-34.)

_____. 1966. *Action and Purpose*. Englewood Cliffs, N. J.: Prentice-Hall, Inc.

Ullmann-Margalit, Edna. 1978. Invisible-Hand Explanations. *Synthese* 39:263-91.

Underhill, G. E. 1904. The Use and Abuse of Final Causes. *Mind* 13:220-41.

Utz, Stephen. 1977. On Teleology and Organisms. *Philosophy of Science* 44:313-20.

Valentine, Elizabeth R. 1988. Teleological Explanations and their Relation to Causal Explanations in Psychology. *Philosophical Psychology* 1:61-68.

Van Parijs, Philippe. 1979. Functional Explanation and the Linguistic Analogy. *Philosophy of the Social Sciences* 9:425-43.

Varela, Francisco, and Humberto Maturana. 1972. Mechanism and Biological Explanation. *Philosophy of Science* 39:378-82.

Varner, Gary E. 1990. Biological Functions and Biological Interests. *The Southern Journal of Philosophy* 28:251-70.

Von Wright, Georg Henrik. 1971. Intentionality and Teleological Explanation. Chap. 3 of *Explanation and Understanding*. Ithaca: Cornell University Press.

Wachbroit, Robert. 1994. Normality as a Biological Concept. *Philosophy of Science* 61:579-91.

Walsh, D. M. 1996. Fitness and Function. *The British Journal for the Philosophy of Science* 47:553-74.

Wassermann, Gerhard D. 1981. Review of *Teleology* by Andrew Woodfield. *Philosophia* 10:125-32.

Wigginton, Eliot, ed. 1972. *The Foxfire Book*. Garden City, N.Y.: Doubleday & Co.

Williams, George C. 1966. *Adaptation and Natural Selection: A Critique of Some Current Evolutionary Thought*. Princeton, N.J.: Princeton University Press.

Williams, Mary B. 1976. The Logical Structure of Functional Explanations in Biology. Vol. 1 of *PSA 1976: Proceedings of the 1976 Biennial Meeting of the Philosophy of Science Association*, ed. Frederick F. Suppe and Peter D. Asquith, 37-46. East Lansing, Mich.: Philosophy of Science Association.

_____. 1981. Is Biology a Different Type of Science? In *Pragmatism and Purpose: Essays Presented to Thomas A. Goudge*, ed. L. W. Sumner, John G. Slater, and Fred Wilson, 278-89. Toronto: University of Toronto Press.

Wimsatt, William C. 1971. Some Problems with the Concept of "Feedback." *PSA 1970: Proceedings of the 1970 Biennial Meeting of the Philosophy of Science Association*, ed. Roger C. Buck and Robert S. Cohen, 241-56. Vol. 8 of *Boston Studies in the Philosophy of*

Science. Dordrecht: D. Reidel Publishing Company, 1971. (Reprinted in *Purposive Behaviour and Teleological Explanations*, ed. Frank George and Les Johnson, New York: Gordon and Breach Science Publishers, 1985, 165-81.)

———. 1972. Teleology and the Logical Structure of Function Statements. *Studies in History and Philosophy of Science* 3:1-80. (Reprinted in *Purposive Behaviour and Teleological Explanations*, ed. Frank George and Les Johnson, New York: Gordon and Breach Science Publishers, 1985, 182-259.)

———. 1976. Reductive Explanation: a Functional Account. *PSA 1974: Proceedings of the 1974 Biennial Meeting of the Philosophy of Science Association*, ed. R. S. Cohen, C. A. Hooker, A. C. Michalos, and J. W. van Evra, 671-710. Vol. 32 of *Boston Studies in the Philosophy of Science.* Dordrecht: D. Reidel Publishing Company.

Winch, Peter. 1975. Comment: Geach on Teleological Explanation. In *Explanation*, ed. Stephen Körner, 95-105. New Haven: Yale University Press.

Woodfield, Andrew. 1973. Darwin, Teleology and Taxonomy. *Philosophy* 48:35-49.

———. 1976. *Teleology.* Cambridge: Cambridge University Press.

———. 1977. Review of *Teleological Explanations* by Larry Wright. *Times Literary Supplement*, June 24, 1977.

———. 1978. Review of *Teleological Explanations* by Larry Wright. *The Philosophical Quarterly* 28:86-88.

———. 1979. Reply to Woolhouse on the Temporal Structure of Goal-Directedness. *The Philosophical Quarterly* 29:65-73. (Reprinted in *Purposive Behaviour and Teleological Explanations*, ed. Frank George and Les Johnson, New York: Gordon and Breach Science Publishers, 1985, 279-90.)

———. 1990. The Emergence of Natural Representations. *Philosophical Topics* 18:187-213.

Woolhouse, R. S. 1979. The Temporal Structure of Goal-Directness. *The Philosophical Quarterly* 29:56-64. (Reprinted in *Purposive Behaviour and Teleological Explanations*, ed. Frank George and Les Johnson, New York: Gordon and Breach Science Publishers, 1985, 268-78.)

Wright, Larry. 1968. The Case against Teleological Reductionism. *The British Journal for the Philosophy of Science* 19:211-23. (Reprinted in *Purposive Behaviour and Teleological Explanations*, ed. Frank George and Les Johnson, New York: Gordon and Breach Science Publishers, 1985, 110-22.)

_____. 1972[a]. A Comment on Ruse's Analysis of Function Statements. *Philosophy of Science* 39:512-14.

_____. 1972[b]. Explanation and Teleology. *Philosophy of Science* 39:204-18. (Reprinted in *Purposive Behaviour and Teleological Explanations*, ed. Frank George and Les Johnson, New York: Gordon and Breach Science Publishers, 1985, 123-40.)

_____. 1972-1973. Teleological Etiologies. *Philosophical Forum* 4:575-84.

_____. 1973[a]. Functions. *The Philosophical Review* 82:139-68. (Reprinted in *Conceptual Issues in Evolutionary Biology*, ed. Elliot Sober, Cambridge: The MIT Press, 1994, 27-47; also in *Topics in the Philosophy of Biology*, ed. Marjorie Grene and Everett Mendelsohn, vol. 27 of *Boston Studies in the Philosophy of Science*, Dordrecht: D. Reidel Publishing Company, 213-42; and in *Purposive Behaviour and Teleological Explanations*, ed. Frank George and Les Johnson, New York: Gordon and Breach Science Publishers, 1985, 141-64.) .

_____. 1973[b]. Rival Explanations. *Mind* 82:497-515.

_____. 1974[a]. Emergency Behavior. *Inquiry* 17:43-47.

_____. 1974[b]. Mechanisms and Purposive Behavior III. *Philosophy of Science* 41:345-60.

_____. 1976[a]. Reply to Grim. *Analysis* 36:156-57.

_____. 1976[b]. *Teleological Explanations: An Etiological Analysis of Goals and Functions*. Berkeley and Los Angeles: University of California Press.

_____. 1977. Review of *Teleology* by Andrew Woodfield. *International Studies in Philosophy* 9:187-89.

_____. 1978. The Ins and Outs of Teleology: A Critical Examination of Woodfield. *Inquiry* 21:223-37.

Wuketits, Franz M. 1980. On the Notion of Teleology in Contemporary Life Sciences. *Dialectica* 34:277-90.

Young, Robert M. 1971. Darwin's Metaphor: Does Nature Select? *The Monist* 55:442-503.

Zumbach, Clark. 1981. Kant's Argument for the Autonomy of Biology. *Nature and System* 3:67-79.

_____. 1983. Is a Genetic Code a Program? The Autonomy of Biology in Ernst Mayr's *The Growth of Biological Thought*. *Nature and System* 5:241-47.

_____. 1984. *The Transcendent Science: Kant's Conception of Biological Methodology*. The Hague: Martinus Nijhoff.

Index